"I love your book....You're like the Energizer Bunny—you just keep truckin' along towards positive change. Your book is very engaging and extremely well spoken. I can't imagine how you go on in the face of such widespread insanity. Yesterday's front page of the Times featured an article about Alaskan coastal villages being wiped out by rising sea levels from melting glaciers, etc. I find it difficult to stay positive when confronted by the political pygmies (with apologies to the real pygmies who've lived in harmony with their forest since God began) who are willing to sell the planet for temporary political gain."

-Jon Cypher, Actor

"Soaring prices, scarcity, and polluted air makes it glaringly clear that our 'energy policy' needs a major overhaul. The most urgent change we must make, as we move toward a sustainable future, is the energy we use to fuel our economy that depends on an energy that is not sustainable. Oil cannot be manufactured; we can't grow it. It is not renewable. Scientists are telling us we have a fifty-year supply left at the present rate of consumption, maybe less. Whatever we will be forced to do then, we should seriously be moving in that direction now. As a scientist, energy expert, and an outside-the-box thinker, Dr. Brian O'Leary is well-qualified to address clean and renewable solutions to the energy and other environmental problems. He expresses them passionately and thoroughly in this book."

-Dennis Weaver, Actor and President,
Institute of Ecolonomics

" I just finished reading Re-Inheriting the Earth and was delighted by it. I thought it was a great survey of where we are now and what we have to do to get where we need to go. Above all, your open-mindedness to the possibilities of new science tempered with a healthy, common sense approach is indicative of the global change in attitudes which is already beginning to take place but needs to become more widespread. It also points to huge possibilities, if only humanity will join together to un-lock that potential."

-John Bunzl, Director
International Simultaneous Policy Org., London, U.K.

Other Books by Brian O'Leary

The Making of an Ex-Astronaut

The Fertile Stars

Spaceship Titanic

Project Space Station

Mars 1999

Exploring Inner and Outer Space

The Second Coming of Science

Miracle in the Void

RE-INHERITING THE EARTH

Awakening to Sustainable Solutions and Greater Truths

Brian O'Leary, Ph.D., copyright 2003
oleary1998@yahoo.com
www.independence.net/oleary

Re-Inheriting the Earth: Awakening to Sustainable Solutions and Greater Truths

Cover and book design by Brian O'Leary Jr.
Cover art by Meredith Miller
Printed on recycled acid-free paper in the U.S.A.

Published by Dr. Brian O'Leary
P.O.Box 27,
Washington, CA 95986.
www.independence.net/oleary

Previously published in Portugese as
Re-Herdar a Terra, 2002, by Ancora
editora, Av. Infante Santo, 52 3.%Esq.,
1350-179 Lisboa, Portugal, 21 395 12 22

Distributed by:

Truth Seeker, Inc.
239 So. Juniper
Escondido, CA 92025
800-551-5328

and

Paradigm Books
P.O. Box 8237
Boulder, CO 80308-1237
970-596-5097

To the Elders

Who have the vision to lead and the courage to change:

Maury Albertson
Alick Bartholomew
John Bockris
Obadiah Harris
Barbara Marx Hubbard
Robert Jahn
Win Lambertson
Bonnie Lange
John Mack
Arne Naess
John Rossner
Ian Stevenson
Nejat Veziroglu
Dennis Weaver
and many others

And in memory of:

David Brower
Yull Brown
Joseph Campbell
Bruce DePalma
Manly Hall
Willis Harman
Shiuji Inomata
Ken Keyes
Bruce and John Klingbiel
Stephan Marinov
Terence McKenna
Donnella Meadows
Sparky Sweet
Swami Vishnu Devananda
Bob Zweig

Acknowledgements

I am deeply indebted to my spiritual partner Meredith Miller who put up with the uneven process of creating this book. She has patiently listened to my tirades about pollution and critiqued my first drafts. I couldn't have done this project without her.

I also thank Bonnie Lange, Susan Ford Collins, Anthony Penna, Brian O'Leary Jr., Father John Rossner, Swami Swaroopananda, Alick Bartholomew, Roger Hill, Peter LaVaute, Obadiah Harris, John Otranto, Ashleea Nielson, Leyla Berg, Karen and Byron Barnes, and Dennis Weaver for valuable feedback and other support along the way.

**From the cover painting *Re-Inheriting the Earth*
by Meredith Miller:**

Emerging into the healing waters of feeling
A man stands alone naked surrendering humbly
His crown of conquests replaced by nature's halo of leaves

He is the true king now, prepared to inherit the
Earth's precious child...himself...

He will re-inherit his own birth of innocence
Into the future Earth of his own making...
May all men remember their vulnerability
Safe in the arms of Spirit Father.

Meredith Miller © 2002

Contents

FOREWORD

After the September 11 Tragedy

TERRORISM, WAR, AND our unsustainable lifestyles add up to a desperate situation affecting all of us. A global governance structure of some sort will become necessary to overrule the systems we now have. We the people need to have jurisdiction over the survival of civilization while preserving the freedoms of individuals. As never before, we global citizens will need to debate a new constitution based on natural law that would remove human ecocide, excessive competition and violence from the equation. The Earth is in the emergency room and is in need of allopathic solutions, such as a solar-hydrogen or new energy economy to replace fossil fuels. We also need to ban weapons in space and limit them on Earth—before both terrorist and opportunistic economic/military threats overwhelm all of us.

Last century, 200 million people died from assaults by weapons and hundreds of millions more were killed by an increasingly toxic environment. These numbers will surely rise this century without massive public participation at a global level. War and tyranny have always been ways of life. But what is unprecedented is that the actions of war and ecological tyranny could finish us all off. We must be called to civil action as a world community.

The latest environmental news is not good. Even the mainstream consortium, Organization for Economic Cooperation's recent Environmental Outlook report (www.oecd.org/env, and www.rachel.org), gives a chilling set of warnings and new statistics, consistent with what will be presented in this book. They describe how bleak the year 2020 would look if we continue to use up our fossil fuels, continue to release large amounts of toxic chemicals into the atmosphere and waterways, continue to ignore the Kyoto agreements on global warming, and continue to deforest, overfish, overgraze, deplete topsoil and water, etc. May this kind of "2020 hindsight" begin to permeate our resolve so that we may avoid catastrophe now, before it's too late.

The American democracy is in extraordinary trouble. At a time when we could be coming together to render humanitarian aid to helpless refugees in Afghanistan and Iraq rather than bombing them, at a time when we must urgently begin the awesome task of restoring the environment, those controlling our destiny have moved in the opposite direction. While some of us are waving flags and closing ranks behind the Bush administration, the voices of reasoned dissent and openness to solutions have become ever more silenced by a tunnel-visioned American media and a narrowing range of debate which could open awareness of other possibilities. The tragic irony of all this is that it need not be that way, that the terrorists could eventually be brought to justice if only we too were to act justly ourselves. We are creating a nightmare of increasing militarism, fear, greed, secrecy, denial, anger, cruelty, pollution, and the curbing of individual freedom of expression. We are also risking a World War III which could end it all.

There are exceptions to the party line, but you'll have to look to find them. Michael Moore's best-selling underground classic *Stupid White Men* (Harper Collins, 2002), and the lucid speeches by U.S. Representative Dennis Kucinich (D-OH) are examples of unreported yet popular efforts to stop the war machine and re-humanize our culture.

U.S. official policies polarize us not only from Islamic extremists but from the rest of the world. I discovered this sobering fact from extensive recent dialoguing with colleagues and audiences in Europe and the daily exposure to the media abroad. As one British elder stated it, "Americans seem to have lost their sense of identity." Perhaps the $60 billion implosion of the Texas energy giant Enron, once the seventh largest corporation in the U.S. and critiqued in this book for its questionable practices in California and its close connections with the White House, symbolizes what might happen when winner-take-all capitalism and exploitation of unsustainable natural resources are allowed to go unfettered.

During the autumn of 2001 I was on a lecture tour of Austria, Germany, Scotland, England, France and California on issues relating to this book. Meredith and I had landed in Paris on September 11 within minutes of the terrorists' horrific attacks on the World Trade Center. Hardly a single European we have encountered out of hundreds of inter-actions supports the frequent American bombing of innocent civilians in poor countries, military tribunals for accused terrorists, the ignoring of international agreements such as the Kyoto protocols on global warming and the Antiballistic Missile Treaty, the accelerating toxicity of our envi-ronment, the ever-increasing hegemony of giant corporations, the arms trade, plans to deploy space weapons, capital punishment, the suppres-sion of new ideas, the media spin, the inequalities of the rich and poor, and the unbridled power of economic globalization, fast-track negotia-tions, with no checks and balances coming from an informed public.

The overwhelming consensus among Europeans is that American-led global cartels involving the abuse of dwindling natural capital, energy, money, food, medicine, government, military and intelli-gence must relinquish their power or the human experiment will have failed.

The common denominator of the current polarization is oil money. Our collective addiction to Middle East petroleum created the powers in charge of both sides of the conflict. We must now act forth-rightly to implement solutions that could end this dependence. As never before, we will need to shine a new light on the world stage, a global democracy/republic whose powers will exceed those who are in charge now, regarding the overarching issue of sustainability. We will need to do all this while upholding the rights and freedoms of the meek—rather than of large corporate interests now in control of our destinies.

I shed tears of joyful sorrow over my own version of patriotism on a recent night of listening to jazz in Paris, music which was first com-posed and performed in America during mid-twentieth century. The spirit of jazz had deeply inspired me during my youth.   I had been a proud Eagle Scout and selected as an Apollo astronaut. I wondered, how can we grasp for those straws that represent the best of us? How can we combine our extraordinary creativity and the blessings of nature we still can enjoy into a sustainable plan? Can we transcend our fears, our work-frenzy, our grief, for just awhile, to embrace our own greatness and let it propel us forward? Only the formation of a firm resolve to move into solutions, and the tests of time, will tell us. All we need to do is to evalu-ate and choose which solutions could lead us into a sustainable future with minimal pain of transition. As one colleague put it "Let's have a positive terra-ist attack and planet together."

There is so much unacknowledged in our culture. There is so much unfinished business and exploring and growing to do. Why do we have to commit homicide, suicide, biocide and ecocide to do our business? We need to recognize the severity of human actions, but we also will need to forgive the transgressions and to accept the situation so that we may end our grief and move into solutions.

We need to develop a new global community in both real and virtual spaces. In his classic book *The Different Drum*, Scott Peck reminds us that community-building includes a chaotic phase which often discourages the founders. This period of ego-posturing usually precedes a surrender to a feeling of emptiness, the next phase. Then it would be possible to enter into the spirit of cooperation and selflessness comprising true community.

Perhaps, in those moments of inspiration, we could turn crisis into opportunity. Perhaps we could express our grief first and then move into responsible and compassionate roles. I hope this book will help shed some light on the solutions themselves. They wait in the wings for their opportunity. In addition, some of us are forming a coalition, which intends to facilitate a new world citizenship which would ensure an enduring civilization through proper human action.

After a death in the family and an unsettled year, Meredith and I have landed on our feet in a place of great natural beauty on the Yuba River near the village of Washington, California. May this New Washington represent the vision for renewed spirit of a peaceful, sustainable and just global future, in sharp contrast to the massive corruption and violence now coming from the old Washington of my past.

Brian O'Leary
Washington, California
November 2002

INTRODUCTION

The New Apollo Program

AS THE NEW millennium dawns, I see hope for us and for reversing the human-caused pollution of the Earth. The changes and adventure will be exciting and challenging for us, yet time is running out. I believe solutions are there, and can be enhanced if we transcend our denial of emerging truths based on suppressed experiments in new science, new energy, healing, consciousness, hemp production, sustainable agriculture and forestry, and evidence for contact with nonhuman intelligence and for our eternal being. I believe we have the potential to make the needed changes, but we are going to have to let go of many worn-out vested interests and begin to empower ourselves toward solutions.

The birth of this process, as almost always, comes from "necessity being the mother of invention." It is clear to me that we as a species must begin to come back into balance with the biosphere. We must create a sustainable future so that we may once again inherit the Earth. We must also transcend the cultural bans on our greater truth whose implications are even more pervasive than those during the Copernican Revolution. As we shall see in this book, not to embark on a new journey towards sustainability and truth will result in a deterioration of the quality of life for our children at best, and global extinction at worst.

As a little boy I always wanted to go into space. There was no space program then. Many of my teachers thought I was just a dreamer. Then Sputnik went up in 1957, when I first entered college. Ten short years later I became an Apollo scientist-astronaut destined to go to Mars. Soon after, NASA cancelled the Mars-exploration part of the program. I went on to other things, but I got to experience the feeling of anticipating what it would be like to go to the red planet. In these troubled times, I am grateful not to have fulfilled that dream in the role of a governmentally constrained spokesman. Instead I'm focussing on a new dream which could not only ensure our survival but open new opportunities which stagger the imagination.

The Apollo Moon program taught me many valuable lessons. I saw it as a crowning human achievement, an example of what we can do as a culture when we put our minds to it. This was an extraordinary, historic achievement — "one step for a man, one giant leap for mankind". To me, the Apollo program epitomized the best of our collective human potential. Ironically, I now find myself embroiled in the midst of a controversy as to whether or not Apollo happened or was a hoax. Fox Television interviewed me about this and quoted me out of context. I gave the impression that Apollo may not have happened, and as a result, I find myself the recipient of dozens of e-mails from supporters and skeptics alike. What an identity to project! To set the record straight, there is no doubt in my mind that the capsules went into orbit around the Moon, because of the photographs, signals received on Earth, and capabilities of the enormous Saturn V rocket booster. It is conceivable but highly doubtful that the lunar landings didn't take place. Who am I to say for sure one way or the other, since I wasn't there? Regardless, the Apollo program was a great success. It gave me a valuable reference point for what we must now do.

Some years later, as a faculty member in the physics department at Princeton University, I began to have some personal experiences that I couldn't explain within the traditional science I was teaching. I was able to psychically tune into a stranger, had a near-death experience, and healed an injured knee through my intention. All this excited me. I began to dream again like the little boy looking through the telescope at Mars and wanting to go there. I questioned the belief that materialism and reductionism were the most general case of our reality. As I stepped ever further out of the cultural box, I grew into understanding some greater truths about reality. This came at a price of losing my visibility and credibility among colleagues.

I asked, "How can the scientific method be applied to studying experiences such as psychic healing, transcendental consciousness, survival after death, communication with nonhuman intelligence, crop circles, and many other denied discoveries that demand unbiased examination?" I was at first surprised to discover that these fields were scientifically further developed than I had imagined, but none of them had yet been integrated into the mainstream. Some of them had, in fact, been denied and discredited. No longer could I hide out in a comfortable scientific specialty. My insatiable quest for truth, beauty and sustainability led to an unanticipated adventure unfettered by our cultural norms.

I began to learn that many basic principles of consciousness and our multi-dimensional being are being confirmed by experiments in quantum mechanics, psychokinesis, alternative healing, clairvoyance and zero point energy generation. I travelled the world visiting the laboratories of unsung inventors and researchers whose discoveries are certain to lay the foundation for the new science. The result of this inquiry led to the publication of my three previous books, *Exploring Inner and Outer Space*, *The Second Coming of Science*, and *Miracle in the Void*.

I am encouraged that more and more scientists are breaking ranks with their materialistic biases to take the courageous stand that our consciousness is the ground of all being. Materialism is but a limiting case of reality and we are on the threshold of a new paradigm which will provide the means for a new renaissance in human affairs. We can learn a lot from what many spiritual leaders and mystics have been talking about for a long time.

My current passion is to release our enormous human potential to balance the Earth's environment. This massive task needn't necessarily force us to immediately adopt radical new technologies before they are thoroughly researched and debated. Free (or new) energy is a dramatic example of what could be done to transcend our polluting ways, but we will need to learn to use this resource wisely. Just in case new energy is abused by military interests, I would further develop low technology solutions as backup, for example, solar, wind and hydrogen power.

I have witnessed and researched a dazzling array of solutions to our global challenges. These will require an increasing awareness of clean technologies, ever further probing our potentials as multisensory beings, and socially inventing those structures and procedures with which to make the necessary changes. We have no other choice.

This is the most difficult book I have ever written. The accelerating pace of ominous current events often overtakes my expression of the situation. It is therefore easy for the specifics to become dated. But the essence of what must be done becomes ever more poignant. In the short space of a few months of writing, we have seen in America the selection of a government firmly dedicated to the interests of big business to the detriment of all life on the planet. And we have seen blacked-out California electricity customers being price-gouged by their energy suppliers who become ever richer and ever more in control of our collective destiny, all masked by a byzantine and distracting utility structure. The only solution perceived by politicians and media is more polluted power.

I prefer to use science as a tool for understanding the physical facts of the matter without resorting to old political and economic theories which underlie our addiction to growth, the consolidation of power, and resulting decisions which favor the few and imperil any sense of implementing solutions. In the process, I hadn't quite expected to adopt a new identity whose expression doesn't match the current paradigms of scholarship, journalism and commerce. Instead, I have become an independent, travelling teacher and student searching for ways to come into balance with our precious environment.

In places, this book may seem to cover obvious ground many of you already know about. This would especially apply to my new kindred colleagues who already understand the depth of the challenge. I'm hoping to attract readers who are either unfamiliar with the gravity of the problem or unknowing about the variety of suppressed solutions that await us. In America, this might include most Democrats, some disillusioned Republicans, young people, and those who didn't vote at all. That's most of us. In other countries, there might be greater resonance with the themes of this book. Some of you who are confirmed capitalists or armed rebels may fall by the wayside. I've made no attempt to manipulate my conclusions to conform to any vested interest or marketing strategy.

My perspective is that of an American dissident and therefore may seem to be parochial to international readers. My reason is that we are the only economic and military superpower, the biggest global polluter but also the crucible of inventiveness and social change. I am familiar with the intense activity of potential change all around me here. Our actions could soon make or break all of us. Commenting on the late but decisive U.S. entry into World War II, Winston Churchill said that America is like a gigantic boiler: it takes time for it to warm up, but once it gets going, there's no stopping it. This pattern could repeat itself.

The United States is the undisputed leading innovator and developer of novel technologies that have changed the world, for better or for worse. In the short order of one century we began the use of electricity, automobiles, airplanes, radios, telephones, television, missiles, bombs, spaceships, lasers and computers which have been aggressively and pervasively implemented all over the world. Whether it be the first flight over the sand dunes of Kitty Hawk or across the Atlantic Ocean or to the Moon, we Americans have always been creators and doers. The U.S. is also the premier place for inventing and researching new energy. But can America turn on the dime of invention and implement solutions, or is the resistance here so great that our friends abroad will need to take the lead? My answer is, both of the above. In the end, we all will need to partner globally on creating new institutions to oversee the needed changes.

We in America inherit a revolutionary tradition of freedom, one which is threatened by environmental neglect. Many of our people enjoy unprecedented abundance. We therefore have an opportunity as never before to empower ourselves into bold resolutions. Do we have the will to join in community and dare to dream of a mega-Apollo project for the Earth? I believe we do, but we need to examine the ways in which we can come together to discuss and debate our millennial solutions. This will require of each of us an inner-sourced empowerment, free of the ego-traps of greed, denial, self-interest and cultural pressure.

Governments will have to change radically, and learn how, at last, to convert swords to plowshares. Private industry will have to change, too, so that the ecology and all humanity will profit from every successful business endeavor. We must re-invent what is meant by the collective global interest, and keep refining it so that our new direction will be felt enthusiastically by all. We could convene a council of elders to form a global green republic, based on an inspired Declaration of Interdependence, to oversee the process of change.

This new project could spin off many others, such as world peace, abundant nutritious food, and the opportunity to evolve to higher states of being as citizens of an eternal universe. Many more of us may soon discover that we are not alone in the cosmos as sentient beings, and that our awareness will never die.

You may not agree with some of the more radical concepts I explore, as they may violate your belief systems. That's fine with me. We all have our beliefs, no matter how sophisticated our intellects may be. Going beyond my tendency as a scientist to be understated and objective, I often use the words "we need to" and "we must" because of the immensity of the crisis before us. Nevertheless, the foundation of this book depends on physical, ecological and scientific truths that no amount of economic, political, academic or media spin could censor. So in spite of the appearance of controversy and a countercultural perspective, I urge you to read on, because the range of solutions is very broad and in the end could be tailored to your evolving world view. The global situation calls for decisive social action far beyond capitalism, communism, modernism and religious fundamentalism. It is time to consider and debate all reasonable solutions. You may have additional ideas not included in this book. They are welcome and I'd like to hear from you. It is time for all of us to meld our dreams.

An Update on the Energy Debate

What a difference a few months make! When I was writing this book late 2000, people seemed oblivious about our energy problems and then made the headlines every day. Soon after, it ebbed under 9-11 only to come back again in ever more exploitation of nature. Yet the press omits any mention of the inevitable demise of the fossil fuel age, which will become necessary because of dwindling supplies and increasing pollution. In the long run, the issue will be joined between two opposing forces. The one now in power will certainly lead our culture into a state of such disarray, there may be no way out. The other offers sustainable solutions.

The American problem is especially acute. The Bush administration has wasted no time offending the rest of the world with its insistence on nullifying the Kyoto agreement on global warming emissions, drilling in the Arctic National Wildlife Refuge, forestalling renewable and clean energy research, building new fossil fuel power plants which would contribute even more to America's lead in carbon emissions, protecting the interests of the rich and big business seemingly bent on war, and promoting a costly and destabilizing missile defense system.

Unfortunately, in the eyes of the mass media, discussion about ending pollution and global climate change has been abandoned. The debate has narrowed to political and short-term economic considerations rather than the long-term physical facts and technological directions which would surely prompt a serious discussion of a sustainable future, as described in this book. Innovation and even talk of innovation are stifled at almost every turn.

An example is a *USA Today* cover article "Six Ways to Combat

Global Warming" (July 16, 2001 issue). No mention is made at all about the real solutions such as solar, wind, hydrogen, improving efficiency and new energy research. Instead, we only heard about political steps such as ratifying the Kyoto agreements (too little too late, but a still an important step towards international cooperation) to nonsteps such as trading emissions credits for the privilege to pollute, doing more studies, or doing nothing at all. How could these be considered ways to *combat* global warming?

Is this neglect of considering the clean, renewable options a reflection of the dumbing-down of America, is it greed, is it fear of the loss of power, money and secrecy? Has the energy cartel joined ideological forces with the media and government to buttress the status quo at any cost? In my more than forty-year history as a senior energy analyst, I have never seen a debate narrow to such nonsolutions which appear as propaganda. Since my childhood, I learned how the USA was a land of invention and opportunity, not an energy-hungry polluting empire which suppresses solutions and is an embarrassment to the rest of the world. This can be very disquieting on the home front and invite ever more denial of our responsibility.

The only global option we have is to cut our toxic emissions to near-zero—a measure which not only prevents planetary suicide but is cost-effective in the long run. Over the coming decades, petroleum and natural gas will become more scarce and expensive, and certainly not cost-competitive with clean renewables. The only beneficiaries will be the giant energy companies, related infrastructure and the very wealthy. The rest of us will suffer from more toxic air, global climate change, escalating prices and more dependence on foreign oil. We can end our addiction to fossil fuels by first acknowledging that they have dominated our international economy and that now we must find ways to replace them through innovation and active civil participation. Once this becomes more widely understood, coal, oil and natural gas will go the way of tobacco, but this time, on a much larger scale and a greater toll of lives. I hope this book will help you understand the depth of the situation and inspire solutions.

We must expand the debate to embrace the real answers. Fortunately, those answers do exist and need to be presented to the public. Our silence in this matter suggests a more insidious aspect of control in the exercise of power. "The most successful tyranny is not the one that uses force to assure uniformity but the one that removes the awareness of other possibilities", said Allan Bloom in *The Closing of the American Mind*.

Noam Chomsky stated the problem this way: "The smart way to keep people passive and obedient is to strictly limit the spectrum of acceptable opinion, but allow very lively debate within that spectrum— even encourage the more critical and dissident views. That gives people the sense there's free thinking going on, while all the time the presuppositions of the system are being reinforced by the limits put on the range of debate."

So what can we do? This book proposes the establishment of a global green republic which would have jurisdiction over unbridled competition and growth of giant multinational corporations and their friends in politically high places. These forces of globalization and "free trade" encourage the very powerful to step into those countries that have the cheapest labor and most relaxed environmental standards in a vicious cycle of pollution and competitive stress. These unregulated actions make a mockery of authentic free trade which could deliver the needed goods and services for a green future.

I will present in Chapter 4 the case for the most urgent measures, ones upon which the preponderant number of citizens of the world would agree. Included would be the shift of public subsidies from polluting to clean enterprises, enforcing strict emissions standards, controlling the excesses of economic globalization, protecting workers, and taxing international currency speculation for the relief of Third World debt and for the creation of new enterprises which would preserve, restore and sustain the biosphere. In order to do this job as quickly as possible, I have become aware of a brilliant idea proposed by John Bunzl in England. It is called the Simultaneous Policy (www.simpol.org).

Under this plan, measures such as those listed in the previous paragraph would be adopted in principle by any individual, organization, city, state or nation that wishes to pledge supporting or voting for adopters. (I am already on board as an adopter.) Meanwhile, business-as-usual would continue to prevail, so adoption would pose no immediate threat to existing policies. By the sheer force of attraction, the Simultaneous Policy would then be adopted by more and more nations, until every country has done so. At that moment, the new measures would be implemented simultaneously by each nation to enact the needed paradigm shift. Ideally, this would be the first step for global green republic in action, without needing yet to establish the new governance structures themselves, which might take a longer time than we have to reverse the accelerating deterioration of life on the planet, as chronicled in this book.

This innovative solution could also allow us plenty of time to debate appropriate measures upon which a majority of people in the world would eventually agree. The idea could be likened to responding to the outbreak of a serious fire. Whereas pouring buckets of tap water on the fire would never do the job (our current situation of legislating incremental solutions), calling the fire department and waiting for their special equipment to arrive would be what the Simultaneous Policy could do: It would give us the chance to put out the fire once and for all, but at a later time. Our global civic responsibility is to call the firemen to come out quickly, at which time decisive action can reverse the damage. We have been asleep at the wheel of democracy while the fire rages on. It is time to awaken to sustainable solutions and have the courage to embrace them.

The process of ratifying the Kyoto agreements on carbon dioxide emissions provides a good example of what could be done with the Simultaneous Policy. It therefore would not be surprising if the United States were to be the last holdout to adopt the policy, but could you imagine the moral pressure which would be mounted against the American Government for not adopting it?

There is a deeper spiritual crisis lurking: current power structures are effectively blocking the need to transform our ethics from ruthless competition to cooperation, from selfishness to selflessness, from separation to unity, from greed to generousness, from staying inside a box of the false security of bread-and-circuses to embracing greater truths of our being. President Abraham Lincoln said: "Nearly all men can stand adversity, but if you want to test a man's character, give him power." Power is vested in the wrong places now, and the necessary shift will need to be returned to the people by a global consensus which has no precedent. The design of a new world governance structure will be one of our greatest challenges and opportunities in human history.

The quest for new knowledge of our being is also a birthright which has been diverted to a mass culture that has arbitrarily separated church and state, has confused dogmatic religion with individual spiritual transformation, and has overlooked the enormous potential of a new science of consciousness. I have noticed that many leaders of the sustainability movement have not gone much further than our political, academic, business and religious leaders have in understanding this point, and so are sometimes missing the full range of solutions and spiritual openings—sometimes called miracles. I'm sure some environmentalists will take issue with my attempts to bridge the gap between sustainability and greater truth. Yet my discussion is well-grounded in collective human experience and scientific study. These paradigm-breaking topics have been omitted in the debate by the same dynamics as the quest for clean energy.

In Chapter 5, I discuss one recent example of this bridge between sustainability and greater truth: the evidence for extraterrestrial life and the UFO phenomenon. This search has only recently gained new credibility under what is called the Disclosure Project, led by Dr. Steven Greer (www.DisclosureProject. org). Over 100 government and military witnesses have testified about UFO and extraterrestrial visitor experiences that add up to a shocking but self-consistent narrative, which goes as follows: After World War II, military facilities in the United States were apparently visited by extraterrestrial beings. This is when the alleged 1947 Roswell, New Mexico UFO crash occurred and the modern wave of UFO encounters began. Perhaps the visits were triggered by America's recent discovery and use of destructive nuclear energy. A massive coverup then began with a growing black-operations which gained and protected for themselves the knowledge of free energy and advanced propulsion. A new secret government formed, in which very few shadowy people knew of the truth, although fragments were known by some individuals, several of whom are only now going public. Even American presidents became excluded from this knowledge, much to the chagrin of President Eisenhower and some of his successors. This rising power from within the government haunted Eisenhower, who warned the people in 1961 in his last speech as president:

"In the councils of Government, we must guard against the acquisition of unwarranted influence, whether sought or unsought, by the Military-Industrial Complex. The potential for the disastrous rise of misplaced power exists, and will persist. We must never let the weight of this combination endanger our liberties or democratic processes. We should take nothing for granted. Only an alert and knowledgeable citizenry can compel the proper meshing of the huge industrial and military machinery of defense with our peaceful methods and goals so that security and liberty may prosper together."

With this statement, corroborated by reputable witnesses, the two topics of sustainability and greater truth become joined. Perhaps we can begin to understand that the genesis of the dominance of fossil fuels and nuclear energy and the suppression of options has been a secret policy of economic and political/military forces whose power Eisenhower felt so compelled to expose. Perhaps we are seeing the advanced phases of a massive conspiracy to profit and pollute by keeping the rest of us ignorant about real solutions. Whether we look at the consolidation of economic power or clandestine political power as the principal cause of our demise, democracy will need a big boost. The next phase of the Disclosure Project would be to have open Congressional hearings. But that can happen only with a civic awakening. Couldn't we all benefit from open inquiry to this mystery? Don't we deserve to know?

Perhaps a significant part of the robust secrecy apparatus has been to suppress "free energy". This is a concept which spans the gap between the sustainability movement and the need to embrace greater truth. As we'll see in Chapter 2, the possibility of free energy is imminent, with hundreds of reputable reports of "over-unity" power coming in from laboratories all over the world. I myself have seen many such demonstrations. Free (or new) energy would cause one of the most extraordinary revolutions in the history of technology (examples include special electromagnetic devices, cold fusion, and hydrogen gas cells). The potential for free energy needs to be debated widely and openly about its great benefits as well as its possible abuse in the wrong hands. Attempts to bring free energy into the world have been thwarted at every turn, because this would spell the end of the energy cartel. As scientist-inventor Dr. Tom Bearden put it, "The cost of a single large power plant will solve the entire world energy crisis forever."

But for understanding free energy and UFO/ET disclosure, many more of us will need to have the courage to awaken to these greater truths, even if they appear bizarre at first blush. Unfortunately our culture has fragmented itself into so many factions that do not communicate with one another, our expanded knowledge and awareness are artificially kept away from us. There are too many vested interests among scientists, industrialists and governmental officials to acknowledge free energy. Yet it exists.

I believe that the awakening will require that we move even beyond the sensible democratic policies of achieving sustainability and curbing the power of multinationals and governments who continue to profit while devastating the planet: We will need to examine who we are, why we're here, what happens after our lives on Earth, and how our con-

sciousness can heal ourselves and the world around us. As we shall see
in Chapters 5 and 6, the clues to answering those questions are everywhere, if only we dare to look.

I know these are radical ideas, and some of you might think them
to be irrelevant speculation. That is o.k. with me. From my experience
and study over three decades, I have learned that going outside the box
of materialistic western science to the realms of consciousness will make
our paradigm shift so much easier. But in the short term, allopathic remedies such as solar, wind, hydrogen and new energy will be absolutely
necessary because the Earth is in the emergency room. We need to lend
all the assistance we can, but we can only do that by a tireless search for
zero emissions, natural preservation, truth and justice.

This book is divided into two parts. Part I describes the tools we
have to restore the resources and beauty of the Earth. Chapter 1 urges
each of us to expand our awareness and to shout our challenges and solutions from the highest mountain, reaching out to our families, communities, churches and the Internet, and (where possible) to the media, academy and corporations. Chapters 2 and 3 describe how we can solve the
energy crisis and restore the biosphere as physical realities rather than as
instruments of fallacious economical and political thinking.

Part II looks at how we can implement solutions in a globally
organized system, what we can do as individuals to meet the challenge,
and the need to expand our knowledge base and establish a new science
of consciousness which will give us the needed foundation for a new paradigm. Think of each chapter as an action item about what we must do.
The chapter titles appropriately summarize these actions.

In my previous book, *Miracle in the Void*, I said that we need not
be afraid to feel our feelings, to grieve the past and to look at transcendent solutions based on our greater essence. The process of grief and
transformation into solutions is an awakening all of humanity will share.
This new action will require nothing less than our most inspiring, inner-
directed compassion and love for all creation. We are at a critical crossroads, and we must act now. It is time to walk through the void to the
miracles that lie ahead, together.

Brian O'Leary
Ridgway, Colorado
August 2001

PART I

Physical and Ecological Action Steps

"The Earth provides for every man's need but not for every man's greed."
-Mahatma Gandhi

PROLOGUE

FEW OF US can doubt that technology is a major driving force behind our contemporary culture. Choices we make in research and development form the thin edge of a wedge that can create multitrillion-dollar industries for decades—and even centuries—to come.

Many of these choices have not been wise. Instead they have escalated towards powerful vested interests in particular directions that can both destroy the Earth and depress the qualities of our lives. In some cases these choices can lead to a mass addiction to near-term profit, which can bring us violence, war, climate change and irreversible pollution.

There is no better example of this imbalance than our consumption of oil, coal and natural gas. Our fossil fuel obsession has such a grip on us that we have become mesmerized about being open to alternatives such as cold fusion and other free energy now being researched, as well as the traditional renewable options: hydrogen, solar photovoltaics, wind power and biomass. One exception to the fossil fuel monopoly is nuclear energy, which has its own problems, such as the disposal of radioactive waste. But we are also seeing other global catastrophes unfolding that demand innovative solutions: water pollution, deforestation, unsustainable agriculture, unreclaimed mining, to mention a few.

Not many of us realize yet that these pressing challenges do have solutions if we simply reprioritize our technologies. Then we can begin to re-inherit the Earth. The following three chapters describe both the current dilemma and state-of-the-art of these new technologies, which are much further along than most of us imagine. In this age of advocacy, commercial distraction, and academic specialization, the truth becomes buried under the simple fact that we must become educated about the full range of solutions based on both low and high technology.

1

Become Aware of the Situation and Be Willing to Look at Solutions

"We are no longer inheriting the Earth from our parents; we are stealing it from our children."

-David Brower

A REVOLUTIONARY WORLD view is emerging. Scores of competent climatologists and ecologists are telling us that human civilization, led by Western industrial interests, may be headed towards global catastrophe if we don't do something very different very soon about our natural environment.

The challenge is, how, when and by whom can this be done? Time and time again, powerful vested interests have carried the day until sensible people seem to have little energy left to resist the onslaught, to think anew, and to guide us out of our dilemma. Like the frog in the pond whose temperature is imperceptibly raised every day to the point where the frog cannot get out, we are stewing in a global juice that is warming and polluting our atmosphere. The worldwide consensus is that the routine burning of fossil fuels is leading us to record heat, drastic climate change, foul air and an uncertain future. We may need to await a major catastrophe to wake us up. Meanwhile, we all share, I believe, a deep unconscious fear of what may come from our senseless binge. We must transcend this paralysis of will in order to make the necessary changes.

The free market system is no longer free. Instead it has been an excuse for giant corporations to control dwindling natural resources without public accountability. This takeover is bleeding the rest of us dry and will ultimately terminate all life on the planet. The market rules are rigged to ensure the success of the controlling elite. They are winning temporarily while the rest of us slip away—unless we join their short-lived gravy train. In the end, we could all lose to the forces of greed.

This process is well-advanced in the energy sector. Some scientists believe that we may not have time to replace our dwindling petroleum supply upon which we depend with any alternative energy source— clean or dirty. The energy policy expressed in the media focusses on continuing our fossil fuel glut as the only way out of shortages. This puts us on a collective suicide course. Meanwhile, the price-gouging by fuel and infrastructure suppliers to California consumers has had the full backing by our elected politicians. These actions do not only pay homage to the awesome power of big business: The suppression of clean and renewable alternatives is a crime against humanity.

There are at least three reasons why we must end the fossil fuel age: (1) supplies of oil and natural gas are dwindling, soon to be half depleted within early this century, (2) burning any carbon compound (especially coal and oil) is highly toxic to the environment, and (3) global warming and climate change are inevitable consequences. Nuclear energy also has major problems. Switching to clean, renewable energy will be the key to a sustainable future. Only when the public begins to exercise its responsibility for energy policy will we be able to start the change.

This one action of taking public control over our energy resources will lead to a new collective responsibility to oversee other natural capital such as food, wood, natural medicines, water, and minerals. The increasing concentration of power of the few over the exploitation of natural resources, and related financial infrastructure which nurtures that power, has kept us away from what we need to do.

We must alter our course away from this tyranny. We need to awaken to reassert our freedom of choice. We must restore democracy to its proper place in public affairs.

The situation is intensified by the fact that, in spite of the appearance of polluted cities of the Third World, the United States continues to lead the way in exploiting the environment. With only five per cent of the global population, we Americans consume one-fourth of the world's energy and one-third of its raw materials [1]. I am not proud of this. My own sense of grief is especially heightened by the fact that I am a citizen of the leading polluter nation, as well as being an individual member of a supposedly sentient species which is causing the greatest mass extinction since that of the dinosaurs 65 million years ago. Most of us are complacent, distracted or conveniently ignorant, in part because of the overwhelming depth of the situation. As Walt Kelly's Pogo has said, "we have met the enemy, and the enemy is us!"

The tyranny we've created about our collective actions is historically unprecedented and will require innovative solutions, some of which I'll be describing in this book. Transcending tyranny is not new to humanity. The U.S. Declaration of Independence arose out of efforts spanning more than a century of struggling to rise above the tyranny of the English king. Of the 56 courageous souls who signed that inspired document, about half were to meet unfortunate fates, including violent deaths and destruction of property. Yet this one initiative formed the glue for founding a new nation, based on the principles of equality, life, liberty, and the pursuit of happiness.

We now encounter a tyranny greater than that facing our forefathers. This time the need is expanded and the source has shifted to globally powerful monied interests who conveniently deny the extent of the problem. While I'm sure the phrase has been used before, we need a Declaration of Interdependence. We are travellers on one planet, one global village, where our own individual and collective actions affect everybody else. This fact must now be acknowledged by the citizens of the United States and of the world. To not do so will seal our fate as we move ever more deeply into uncertainty about our future.

The currently adopted piecemeal approaches to the challenge show that what we've done so far is not working. Virtually all our existing institutions are guarding their own self-interest and have become part of the problem rather than of the solution. By their actions they perpetuate the myth that the economy transcends the ecology. Few realize that you can't have an economy without a global life support system.

I'm convinced that we need an environmental Bill of Rights, and it can only come from the people. Every great journey starts with small steps. In 1776, a small percentage of the public supported the Declaration of Independence, yet that action was sufficient to expand a revolution leading to the founding of an enduring, albeit flawed democracy. We can only hope the coming revolution will not be violent and that we can all work on this together. The task will be enormous yet rewarding: we'll return to this in Chapters 4, 7, and 8.

The literature on the past, present and near term future state of the world is well-researched, factual and sobering. No amount of political or economic posturing can mitigate the physical facts. I have recently read thirty or so thoughtful books about the gathering storm. Some extraordinary insights await those who seek the knowledge.

The Situation is Not New, but It's Getting Far Worse

Many students of ecology cite the 1964 release of Rachel Carson's classic *Silent Spring*[2] as the beginning of the modern period of environmental awareness. Her warnings about the threat of DDT and other pesticides on the survival of other species and on our own public health were a masterpiece of investigative reporting. Her work triggered modern ecological thinking during a time in which we were more open to new ideas. I believe our situation now is far more serious. We must wake up to that kind of openness that greeted Carson's work once again.

Going back even further was conservationist William Vogt's prophetic book *Road to Survival*[3]. Written in 1948, this work reads more like an inspired contemporary piece about the environment than an outdated protest. Citing statistics such as the loss of one-third of American topsoil and one-half of its forests over the previous 150 years, Vogt lamented:

"If we are to make peace with the forces of the earth, that peace must begin in our minds—and we must seek, and accept, many new ideas. We must reject many old ones...one of the strangest lacunae (lacks) in human cultural development is the absence of understanding of man's relationship with his physical environment. So anthropocentric has he been that, since he began to achieve what we call civilization, he has assumed that he lives in a sort of vacuum."(p. 47)

Vogt laid the blame squarely on industrializing ourselves and our landscapes without limit. "Industrialization", he wrote, "(made) it possible during a hundred years for the most powerful sector of the human race to live as though it were independent of the earth...The methods of purchasing power go back, fundamentally, to natural resources—especially the land—and no amount of symbolical juggling of 'capital' can help us escape that hard fact. There are too many people in the world for its limited resources to provide a high standard of living. By use of the machine, by exploitation of the world's resources on a purely extractive basis, we have postponed the meeting at the ecological judgment seat. The handwriting on the wall of five continents now tells us that the

Day of Judgment is at hand."(p.78).

"The methods of free competition and the application of the profit motive have been disastrous to the land...Business has been turned loose to poison thousands of streams and rivers with industrial wastes..."(p.34).

The late forties: this for me was a time of youthful revelation about the wisdom and beauty of nature. I recall the summers in New Hampshire's then-pristine Lake Winnipesaukee, birch trees, mountains, canoeing, punctuated by the occasional outboard motor sounds ripping through the air, portending a noisy future. Otherwise I grew up in suburban Boston with its oppressive summer heat, its black spring snow, the bright new 1949 Pontiac in the garage, the maple climbing trees, basketball, baseball, tennis, track, piano, a fascination with large numbers, secret crushes on girls, famous Harvard professor neighbors, air raid shelters, the fear of nuclear attack by the Communists, the driven work-ethic of my father and teachers——and a refreshing encouragement to be intellectually creative, drawing on a rich imagination which I was able to express, often to the amazement or chagrin of my parents and older siblings.

Overriding these youthful impressions was a single-minded purpose to my life: I wanted to explore space. First motivated by awe, and then as an adult by scholarship and economics, I led efforts to establish a new human beachhead free of the industrial frenzy of Earth. After getting a Ph.D. in astronomy at the University of California in Berkeley, I became an astronaut and an Ivy League professor as a planetary scientist. Several years later I joined the physics faculty at Princeton University, where I worked with space visionary Gerard O'Neill on industrializing space. We discovered that this could be done cost-effectively if we use the resources of the Moon, asteroids and moons of Mars. Their light gravities and the availability of full-time solar energy in high orbits would allow the human economy to expand far beyond its terrestrial limits. Remarkably, we could do all this with today's technology. But from my newer perspective as an ecologist, I now have second thoughts of expanding our polluting enterprises beyond Earth while we keep soiling our own nest. The post-industrial exploration of space could become appealing once we have created sustainability here and have developed advanced technologies such as anti-gravity. But now we spend billions of dollars for a lackluster space program whose potential seems to serve only an elite: "Scarsdales in orbit", as anthropologist Margaret Mead once put it. (Scarsdale is an elite New York suburb.)

I wonder what would have happened if I had read Vogt's book as a young person rather than now. I instead had chosen the path of space by reading everything I could get ahold of in space exploration concepts, from Werner von Braun's visions in *Collier's* magazine to Arthur C. Clarke's writings about space satellites. Nearly everybody thought I was crazy to harbor such futuristic ideas, because there was no space exploration program then. Of course, I turned out to be vindicated on that one. What if my career path were instead to have been an ecological one? It certainly is now as I move into my sixties, a late career shift. But as this book will show, I trust technology more than many traditional ecologists such as E.F. Schumacher, Paul Ehrlich and Amory Lovins, who turn away from new energy and other novel sustainable concepts that could be compatible with our expanding population. These technologies are hidden from view yet promise a clean, renewable and affordable future.

For all the years we've been in space, there is little we have to show for it in terms of cleaning up our messes. Besides some passive symbols such as the photographs of the Earth taken by the astronauts revealing the awesome beauty of our fragile biosphere, we also have the satellite photographs that remind us about how we've been messing things up, ranging from the human-caused eroded silt fanning out from the mouths of our great rivers to the systematic destruction of the waterways, forests, grasslands, croplands and coral reefs. But we need to go beyond monitoring the obvious: we must *do* something about it.

More than fifty years later the facts show that Vogt's warnings can be amplified to a shrill alarm. I wonder how his spirit would now feel if he were to read the statistics of the extent to which our global life support systems have been depleted between the middle and end of the twentieth century. During that interval, another third of the world's remaining forests have been wiped out. As of 2000, eighty percent of the standing two-thirds are endangered. Why? Because of land-clearing practices and the demand for paper, up seven times since 1950, lumber up two times and fuelwood up three times.

Worldwide energy consumption, mostly of the polluting kind, has quadrupled since 1950. In the U.S. alone, we use as much energy today as the whole world consumed during that year! Globally we have ten times the number of automobiles we had back then. Water use is up three times since midcentury, with water tables falling rapidly. Most river water has already been siphoned off for human consumption as a result of a threefold increase in irrigation. Rangeland productivity has levelled off after a tripling, thanks to diverted water. Fertilizer use has gone up nine times as ever more topsoil becomes chemically polluted and eroded away in our agricultural and pasture lands. The fish catch has quintupled, with the stock dropping fast. Not the least disturbing is that the world population has more than doubled since 1950, and continues growing at the rate of ninety million people each year. There is little left to feed that population as the agricultural bag of tricks becomes exhausted.[1]

Vogt speculated about what might befall us in his future, which is now, if we were not to change our ways. "What is (our security) going to be worth forty or fifty years from now", he wrote, "when millions really need it, if our soil and water, forests and grasslands, the basis of our national wealth, have been destroyed at the same rate as during the past fifty years?"(p.150) Of course we haven't changed our ways. If he were still alive Vogt would think we are truly fooling with creating a Sword of Damacles hanging over us in reprisal for our actions. The process of biospheric destruction is well advanced.

Global warming and human-caused climate change were unknown to Vogt, although he predicted that dust bowls would form during coming climate instabilities over lands we have already eroded. That indeed is happening at an alarming rate all over the world. He just didn't know that the widespread climate instabilities themselves may be human-caused, giving desertification an added boost.

The dramatic increases in the pollution of the air and water, along with the toxic and radioactive waste, have already caused more death and destruction than Mr. Vogt could ever have imagined in his worst nightmares. True, the hunger challenge has not spread as much as it could have, although starvation and famine continue at an intolerable rate. Over one billion people continue to live in abject poverty with little available food. Regarding world hunger, some time has been bought in terms of agricultural yields from the Green Revolution and various developments in biotechnology. But many of these "improvements" have a high price, including the narrowing of crop diversity, allergic reactions, nutritional uncertainties and the farmers' increasing dependence on large corporations for their livelihoods. These new seeds produce crops that do not increase productivity, except to resist pests. For these

reasons some European countries have wisely banned genetically modified foods until we understand their true impact on the environment and
our bodies.

The numbers speak for themselves, as we begin to exceed more
and more limits to sustainability. We've somehow lost our common sense
about our relationship with the natural environment and therefore our
collective future. The twentieth century was our most violent and heed-
less one, and continues to escalate into the current century. The first half
of the twentieth century revealed massive death by the war machines of
two world wars; the second half vindicates Vogt's perceptive remarks
about our declining environment. We are going to have to rethink, and
I know Mr. Vogt would agree: Something *must* be done.

The Big Fossil Fuel Burnup

At the top of the list is how we abuse our energy resources. We
are consuming oil as if there were no tomorrow. We in America are particularly culpable, as half of U.S. car sales at this writing are sports utility vehicles that typically burn up a gallon of gas to go ten miles. Speed
limits on the open highways are back up to seventy miles per hour and
are often even higher. "Americans' energy-intensive lifestyles, and the
U.S.-led global energy consumption trend of the past century—a 10-fold
increase and a quadrupling since 1950—cannot possibly be a sustainable
model for a population of more than 9 billion in the twenty-first century",
say Christopher Flavin and Seth Dunn of the Worldwatch Institute [4].

And around 2050, we will have depleted most of the Earth's oil,
almost all within a century of greed and callousness. Most experts agree
that soon oil production will begin to decline as half the reserves will
have been used up, with sharp rises in price. Here we have a natural
resource painstakingly formed by the natural decay of living matter over
the eons suddenly being wrung out of the biosphere in a time span millions of times shorter. Are we crazy?

The world's first and only trillion-dollar-per-year industry was
founded on fossil fuels. Through well-funded public relations campaigns, this almost unlimited vested power downplays the high cost of
air pollution, acid rain, oil spills, oil wars, unsightly grids and power
plants, and global climate change. To the contrary, thousands of mainstream scientists agree that this "second-hand smoke" from incinerating
oil, coal and natural gas is causing more death and suffering than tobac-

co could have ever provided. We are all in this one together.

Just as alarming is that climate change almost certainly comes from our routine burning of fossil fuels through global warming. We now have more carbon dioxide in the atmosphere than we've had for 160,000 years. Global temperatures are definitely rising, with each successive year becoming markedly warmer. Weather disaster relief costs have increased tenfold over the last thirty years of the twentieth century, totaling a half trillion dollars. These bills continue to climb ten per cent each year, bringing the world ever closer to bankruptcy. The decade of the 1990s has been the hottest in recorded history. The results already include unprecedented liquid water at the North Pole, the breaking off of the Antarctic ice sheet, the melting of ice caps, glaciers and permafrost, the rise of sea level, the erosion of beaches, destroyed coral reefs, heat waves, dust bowls, forest fires, floods, mudslides, super-hurricanes, super-tornadoes, and the substantially heightened breeding and spreading of airborne diseases.

By far the warmest year in historical times was 1998, a record that could go back more than 600 years. Worldwatch has estimated that 32,000 people died in 1998 from weather disturbances, and 300 million more were driven from their homes. Just as ominous is the rate of warming that year, one of the largest on record. These spurts during the 1990s has made the twentieth century the warmest in 1200 years.

Global warming is exacerbated by logging and agricultural land clearing practices which are decimating the tropical rain forests in particular. These activities significantly reduce the effectiveness of plants to absorb the excess carbon dioxide in the atmosphere, and to produce as much cooling rainfall.

The "greenhouse skeptics" paid by oil and coal interests would want you to believe otherwise: that this may be just a coincidence and that climate change could be a natural process. We don't have proof yet, they reason, so therefore we can continue to pollute. But, based on greenhouse models and resulting climate instability, 2500 relevant scientists comprising the United Nations Intergovernmental Panel on Climate Change (IPCC) in 1995 concluded, "A pattern of climactic response to human activities is identifiable in the climatological record."[5] During 2000 the IPCC scientists revised their models and made their point even stronger: "there is discernable human influence" on our climate, they said. They raised their estimates of global warming to an average temperature at the Earth's surface of between 2.7 and 11 degrees Fahrenheit by 2100.[6] This nearly doubles their previous projection.

Their reason for the increase in global warming projections for the coming decades is an interesting study of ecological feedback loops— this time when efforts to clean up one subsystem can adversely affect another. The IPCC's warmer revision comes from the reduction of sulfate emissions from factories and coal-fired power plants that block sunlight and cool the atmosphere. This once compensated for global warming significantly.

Things could get even worse. A special issue of *The Ecologist* points out insidious feedback loops that could trigger a runaway greenhouse effect. The human destruction of trees on land and phytoplankton in the oceans is reducing the Earth's capacity to absorb carbon dioxide from the atmosphere.[7] Another factor is that methane is outgassing from the Arctic permafrost at accelerating rates while the warmer temperatures are melting and vaporizing the permafrost. Methane is twenty times stronger than the same quantity of carbon dioxide as a greenhouse gas. The icing on the warming cake is that our gaping polar ozone holes allow more sunlight to come into the atmosphere and be trapped as heat.[8]

But the upward spiral of temperatures may not end here. Rising levels of oxides of nitrogen (NOX) from fertilizers dumped into coastal waters add even more greenhouse gas to the atmosphere.[9] While the quantities are smaller, the NOX gases are 200 times stronger in warming the air than the same amount of carbon dioxide. They are also poisonous.

The IPCC researchers did not look at these things in their models, so we can fully expect even more drastic global warming and climate change than their models predict. Gloomiest of all is that atmospheric scientists have calculated it would take decades to bring greenhouse gas levels to their natural values, even if we were to magically end our fossil fuel binge tomorrow.

"Our climate is now changing," said James Baker, undersecretary of the U.S. National Oceanic and Atmospheric Administration (NOAA) and Peter Ewins, head of the British Meteorological Office (BMO) in a joint statement. [10] "Ignoring climate change will surely be the most costly of all possible choices, for us and our children." It is wrong to imply, they asserted, that concern about global warming is being exaggerated.

Another team of scientists from NOAA gave a 90 per cent probability that extreme weather we are now experiencing worldwide is due to human activities. At the top of their list of causes was the dumping of carbon dioxide and other greenhouse gases into the atmosphere, almost entirely attributable to our consumption of oil and coal[5] and to unsustainable logging and slash-and-burn land clearing[11].

Still another group of Earth and atmospheric scientists comprising the American Geophysical Union (AGU) issued a January 1999 policy statement that there was a "compelling basis for legitimate public concern" about human-induced climate change. Scientific uncertainty, they said, "does not justify inaction." They also warned that there is no geological precedent for the sudden increases of carbon and other greenhouse gases from the burning of fossil fuels. As former secretary of the planetology section of the AGU, I can personally attest to the scientific rigor and diversity of this professional society regarding the terrestrial sciences.

In fact, the four principal organizations of relevant mainstream scientists (IPCC, NOAA, BMO and AGU) have unanimously agreed that we have created for ourselves a world of unprecedented climate change. In October 1998 yet another group of leading scientists declared in the prestigious journal *Nature* that global climate change could soon become the "environmental equivalent of the cold war", and that these problems need to be addressed with the urgency of the Apollo program.[4] And yet we continue to hear denials coming from the greenhouse skeptics.

Journalist and author Ross Gelbspan posed the situation this way: "The (skeptics') campaign has exerted a narcotic effect on the American public. It has lulled people into a deep apathy about the crisis by persuading them that the issue of climate change is terminally stuck in scientific uncertainty. It is not."[5].

"When the nations of the world convened in Kyoto, Japan in late 1997 to negotiate an international climate treaty—with emission reduction levels and timetables—they carefully calculated the costs of such reductions to their own industries, the possible shifts in their economic relationships with other countries, and the requirements of their domestic political constituencies. Meanwhile Antarctica is fracturing. The ocean off California is becoming a wasteland. Plants are migrating up mountains to keep pace with the warming climate. Whole species are migrating under the surface waters of the seas. The Arctic soil is warming. The oceans are rising. Tropical diseases are moving north. High in the mountains, the glaciers are shriveling. Forests are rapidly losing their ability to thrive. And insects are poised for a population explosion that would jeopardize our food crops, our trees, and our health."(p.151)

The greenhouse skeptics also brush aside the enormous implications of rising the mean global surface temperature by just a degree or
two. "The climate system", according to Worldwatch, "is nonlinear and
has in the past switched abruptly—even in the space of a few decades—
to another equilibrium after crossing a temperature threshold. Such
shifts have the potential to greatly disrupt both the natural world and
human society. Indeed, previous changes have coincided with the collapse of several ancient civilizations."[4]

In spite of all this, the greenhouse skeptics still carry on. One of
the most vocal is American physicist Fred Singer.[11] We were close colleagues during the 1970s and 1980s when we worked together to find cost
effective ways to access the surface of Mars by way of its moons Phobos
and Deimos.[12] But we are now sadly divided by this issue in which he
strongly disagrees with the overwhelming consensus among thousands
of atmospheric scientists that humans are causing most of the global
warming and climate change. The IPCC carbon emissions and warming
models correlate well with the actual surface temperature increases during the twentieth century. The fact is, the fossil fuel industry supports
Singer and his colleagues, who comprise a shrinking minority. For thirty years I was an active planetary scientist and can vouch for the fact that
the professional work of the IPCC and others is sound. I believe the work
of the skeptics is deeply flawed.

Even if we North Americans, Europeans and Japanese were to
reform our polluting ways, the situation in Asia is not encouraging. With
its population five times greater than that of the U.S. and an ambitious
industrial future, China is now second only to America in producing
greenhouse gases. Along with India, whose growing population also
exceeds a billion, the two Asian giants are increasing their emissions at
the world's fastest rates. And global warming is just part of the equation.
Estimates published by the World Health Organization and World Bank
in late 1997 show that well over two million people in China and the rest
of Asia die each year from the effects of air and water pollution—more
than died during the entire Vietnam war.

We see little talk about this in the media. By focusing instead on
specific disasters, they explore only effects rather than probable causes.
They rarely discuss solutions and they underplay humankind's influence. For we as a culture are in denial and apathy about our responsibility in creating an unstable future.

What is Our Future?

Human-induced climate change, global warming, water and air pollution, growing plagues, mass extinctions, deforestation, desertification, unmitigated growth, urban sprawl and economic greed combine to form insidiously interlocking feedback loops which bring us ever closer to our hour of reckoning. In his well-researched recent book, *The Future in Plain Sight*, author and *Time* correspondent Eugene Linden [13] writes, "Climate instability, should it continue or worsen, will further increase the likelihood of a (political or economic) crash and could exacerbate volatility in food supplies...What does climate instability portend for the food supply as the population continues its trajectory to eight to ten billion souls?" (p. 110)

"When ecosystems are out of balance, microbes tend to benefit, when populations of any given species explode, disease can bring them back into balance with brutal efficiency." (p.111)

Linden referred to the publication in *Science* in 1997 of an account by a team of ecologists led by Peter Vitousek of Stanford University. "The figures this group produced are awesome: half the world's mangroves, vital buffers and nurseries of the oceans, altered or destroyed; 66 percent of all recognized marine fisheries either at the limit of their exploitation or already overexploited; half the accessible fresh water on earth co-opted for human use; roughly one-quarter of all bird species on earth driven into extinction; and on and on." (p.100)

"Lurking in the future", Linden continues, "are the unfolding consequences of ozone depletion, which may be weakening the immune systems of many creatures on the planet, and the dislocations of ecosystems that may come from climate change." The alarming disappearance of ozone in the stratosphere protecting us from lethal ultraviolet radiation comes from human emitted chemicals that also reinforce global warming itself.

One doesn't have to be a rocket scientist or doomsdayer to believe that if we don't radically change the way we do things in the new century the "end times" could be upon us—the sinking of coastal cities, plagues, mega-storms and so forth. Authors Peter Russell [14] and Terence McKenna [15] foresee an inevitable crescendo during the twenty-first century because of the accelerating rate of human impact. Analysts of chaos theory call this a phase change—that moment when a system previously out of equilibrium transforms into something radically new.

McKenna's time lines for this change point to around the year 2012, which also happens to mark the end of the Mayan calendar. He sees this as a sort of big bang in reverse, an inevitable event towards which we are gaining speed, with unknown consequences. Like a train out of control this race to omega point can feel breathless to many of us. It is hardly empowering for us to idly watch each accelerating event that has an escalating life of its own, with a dimming hope for reversal. With little to say about it, we seem to speed up our own pace of life, in a vain hope for expanded opportunity, and those chances appear to shrink.   Yet as a species we are the ones responsible for the escalation itself, and are therefore obligated to lead the way out. We need a new consciousness and all the help we can get. Einstein's famous statement "No problem can be solved from the consciousness that created it" suggests that non-humans may need to assist us. We'll explore that question in Chapter 5.

In a league of its own, much of the rest of what is called "futurology" seems to be an armchair exercise in escapism. Most of these people ignore the possibility that our ecological challenge is very real and that the paradigm-busters such as new energy research could radically change our world view. In these nonvisionary times, the only future envisioned by the Disney Corporation, for example, are "retrospective futures", where we look at the coming world from a nostagic, well-animated nineteenth century Jules Verne perspective. Based on some recent interactions with their executive team, I discovered that the most radical concepts they could come up with for the new century were the latest electronic gadgets, rockets to Mars, web TV and a computer in every classroom. This is insignificant in light of what's happening.

"Today's futurists", wrote Worldwatch, "look ahead from a narrow perspective—one that ignores some of the most important trends now shaping our world. And in their fascination with the information age that is increasingly prominent in the global economy, many observers seem to have forgotten that our modern civilization, like its forerunners, is totally dependent on its ecological foundations."[1]

Whatever we might want to believe about the future, it is clear we are altering our environment in ways that portend an unstable destiny. All this need not be so.

The Deepening Crisis and American Responsibility

Our overburning of oil and coal is but one aspect of the escalating ecological crisis which will need immediate addressing. We must also reverse the slashing and burning of our forests, the lowering of water tables, the dwindling sizes of fisheries, the destruction of croplands and rangelands, the lethal and wasteful actions of war, the ozone depletion and the most massive species extinction since the time of the dinosaurs. We have dropped dangerously below the threshold of sustainability for many systems, and the price of restoration is already high, and will get exponentially higher with each day we don't act.

The cost of restoring our ecological systems is unknown, but will certainly run into trillions of dollars per year. It depends on who we ask and what our standards will be, a matter of enlightened public debate. Of one thing I can be sure: this effort will involve more participation than any single initiative in the history of the world. It will be humanity's first true global public project and must be properly managed first by a robust and reformed United Nations and then by new governmental structures I'll suggest in Chapter 4. Global capitalism is not nearly adequate to the task and in fact suppresses the solutions. We will need to come together as a new team.

We Americans, instead of reversing our conspicuous consumption of unrenewable resources, continue to be hooked on gobbling ever more while maintaining tens of thousands of weapons of mass destruction even though the Cold War is over. We carefully police the status quo while some of the brightest new ideas are swept under the rug or are kept from us within the bowels of unknown control groups and secret agencies. The United States now leads the way in domestic imprisonment rates and in unleashing its war machine abroad, with no end of escalation in sight. Why can't we reallocate our military spending to clean up the environment and have a UN police force handle the problems of a Bin Laden or Saddam Hussein?

The American reputation abroad is not a good one. In November 2000 the industrialized nations met in the Hague, Netherlands to ratify the 1997 Kyoto Accords on greenhouse emission reductions. The conference ended in spectacular failure because the United States refused to fulfill its own Kyoto pledge. Instead our representatives to the conference said we would rather continue polluting the atmosphere with larger and more vehicles and powerplants in exchange for paying off emission credits (read: the right to pollute more and pay something for the privilege) and planting ugly monocultured forests to absorb a token trace of carbon dioxide from the atmosphere. I'm not proud to say we would rather do little or nothing to cut the emissions themselves, as called for by the Kyoto Protocols. The Bush administration has since opted out of the Kyoto agreements entirely, reversing its own promise to curb emissions from power plants.

The Kyoto cutbacks would be modest compared to what needs to be done to stabilize climate. While the industrialized countries outside the U.S. have promised to cut greenhouse emissions by a 5-50 per cent less than their 1990 levels by 2010, the reduction would really need to be more like 80 per cent [16] to bring down global warming to acceptable levels in light of the IPCC's most recent projections. "So everyone went home mad (at us)," said the late Donnella Meadows, adjunct professor at Dartmouth College and director of the Sustainability Institute in the U.S.[16]. The anger was especially poignant when delegates at the Hague heard from representatives of the low-lying atoll nations of the South Pacific who face the permanent inundation of their land, forcing them to evacuate because of sea level rises and El Nino storm surges. "After eight years with (former Vice President) Al Gore in as much power as he may ever be, " writes Meadows, "our country is far from a global leader on the issue. We are the obstructionist, the outlaw, the Saddam Hussein. And George W. (Bush) cares as much about climate change as you would expect from a Texas oilman."

Meadows argues that we don't need to wait for an enlightened U.S. President to lead the way out of climate change. "Whatever the U.S. does, Denmark, the Netherlands and Germany have detailed plans to cut their greenhouse emissions by 20 to 50 percent—and in the process pioneer and patent the new energy technologies that will inevitably replace coal and oil." These new energy solutions form the major thesis of this book. We'll explore them in the next chapter.

We can find symptoms of the American problem in our media— for example, the special Earth Day 2000 issue of *Time* Magazine entitled "How to Save the Earth." While the articles themselves were of high quality as in the case of the special issue in Britain of *The Ecologist* on climate change, Time went commercial. Included were 36 pages of advertisements from the Ford Motor Company touting "better ideas" and thinking "outside the box" about their own tiny, incremental improvements in emissions which are more than offset by their manufacturing of gigantic gas-guzzling sports utility vehicles (SUVs). That fact was hidden from the readers, of course; there were no SUV advertisements in that particular issue.

I don't mean to imply that only we Americans are culpable and that the Brits are exemplary environmentalists eager to get into clean and renewable energy. In one report, "While most members of the European Union generate 10 per cent of their energy needs from renewable sources, on average, England totals less than three per cent." [17] For example, while Germany plans to have 100,000 solar-powered buildings by 2010, the United Kingdom plans to build only 100. Most of England's renewable energy comes from hydroelectric dams which often destroy the landscape and wildlife.

"Britain is reportedly on track to meet its commitment of 12.5 per cent" reduction in carbon emissions, says Zac Goldsmith, editor of *The Ecologist*, "but our government still refuses to take real action to combat the problem...Instead, Tony Blair has promised to spend ten times less on renewable energy investments—necessary, he has said, to avoid catastrophe—than he spent on a plastic dome." [18] (The massive Millennium Dome in London is considered by many to be a political boondoggle).

But the American problem looms greatest. Regarding the Hague conference, Goldsmith said, "The Americans, in particular, refused point blank to play ball. Surrounded by the usual lobby-Mafia—nuclear, logging and oil representatives, as well as armies of clever dick lawyers—the country that produces a quarter of all global emissions with just five per cent of the world's population defended the big business line...But the truth is, while appearing to comply, we will in effect see a net increase in US emissions—by roughly the same amount they were supposed to decrease their emissions. This they have done expertly by insisting on the expansion of endless loopholes in the agreement."

"Economic losses from natural disasters," added Goldsmith, "increased from $53 billion in the 1960s to $480 billion in the 1990s according to the German insurance company Munich Re. About 80 per cent of these costs resulted from extreme weather."

Meanwhile we Americans are peppered daily with "environmental" ads from petroleum, coal, utility and auto companies. I don't mean to totally dismiss all these powerful industries and their politician partners for their eco-propaganda (ecopornography?), because alas, we may some day need to work together to implement the needed changes. It's ironic that we Americans don't only control the seeds of global destruction; we also have the potential of leading the way into solutions. As I'll describe in the next chapter, we have the inventive genius and vast resources to do the job. But we will need a paradigm shift. Sadly, our nation is increasingly policing the global status quo while neglecting the basic needs of its citizens. During the last twenty years, the incarceration rate for nonviolent crimes in the U.S. has tripled to 1.2 million. As governor of Texas, George W. Bush has overseen more executions than in any state at any time. As U.S. president, Mr. Bush has revoked any sensible policy on the environment, labor, justice and compassion for life.

We Americans have no universal health care, outmoded election laws and a Supreme Court which can now exclude from office winners of popular votes in close elections such as the 2000 Bush-Gore fiasco. Whatever happened to the days when we were the good guys in World War II? Why do we have more than 35,000 firearm murders per year versus just a handful in other nations? Why the school shootings? Do our children sense a lack of hope coming from our leaders about the kind of world they will be inheriting? Things seem a lot bleaker than when I was growing up. Have we become too rich, smug and superior for our own good?

Perhaps much of this malaise can be traced to the fact that U.S.-led global and domestic actions continue to be non-visionary and near-term business-as-usual. Policies seem to depend on quarterly bottom lines, terms of elective political office, media gossip, hidden agendas, and the threat of war and natural catastrophes. By its inaction, the U.S. Congress has become calcified by special interests. The quest for profit has overwhelmed our ability to cope with the situation. President Bush and Vice President Dick Cheney are oilmen who prefer voluntary industrial emission controls over the law. Following their lead is like asking for the foxes to guard the chicken coop. Thanks to their inside connections, they collected more money for their campaign than in any previous election in American political history. In these times of one dollar, one vote, it should come as no surprise that a large number of their donors represent the very lobby groups in the fossil fuel, utility, automotive, pharmaceutical and chemical industries and their legal and financial teams whose favors are now being returned. Vice President Cheney, former CEO of Haliburton, the largest oil equipment company in the world, received a "golden parachute" retirement award of $20 million so that he could return to "serve the public."

The Bush administration reflects the culmination of the pollution paradigm. History seems to be repeating itself and Bush appears as a perfect mirror of his father, the American President between 1989 and 1993 and before that as CIA director. At this writing, the oil-motivated

Gulf War team gathers again. In censoring the 1989 Senate testimony of NASA climatologist James Hansen on global warming and climate change, the senior Bush had shown his true identity as pro-industry and anti-environment. During the 2000 political campaign, son George W. Bush's defeated Democratic opponent Al Gore joined the forces of denial by his own silence on the objectives stated in his book *Earth in the Balance*. One of his goals was to eliminate the internal combustion engine by 2020. In 1992, former President Bush said this about Al Gore: "This guy is so far off in the environmental extreme, we'll be up to our neck in owls and out of work for every American. This guy is crazy. He's way out, far out, man!" (Applause) Yet even Gore himself has sold out.

At the turn of the millennium, U.S. government subsidies (corporate welfare) directed towards coal, petroleum and diesel research exceeded $20 billion each year. Little has been allocated to clean and renewable energy. Why must we stand by helplessly and watch this selfish lobbying take over our lives? Couldn't we expand our choices? What has happened to our democracy of one person, one vote? We must be aware that the political and social obstacles to change are far greater than the perceived technological and economic obstacles.

From Economic Elitism to Ecological Realism

Not only are politicians being bought out, but the inequality in wealth among the world's people keeps growing and the U.S. once again leads the way. "It is now becoming obvious", says Worldwatch, "that the widening gap between rich and poor is untenable in a world where resources are shared. In the absence of a concerted effort by the wealthy to address the problems of poverty and deprivation, building a sustainable future may not be possible."[1]

J. Brian Atwood, for six years the head of the U.S. Agency for International Development (AID), stepped down in June 1999 in protest of U.S. Third World policies. The industrial world is getting "shamelessly rich" while most of the people are losing ground, Atwood said, putting the ratio of earnings in rich countries to those in poor countries at 65 to 1. He blames American political leadership for holding back on funds for health, education, job creation, community development and food security. The U.S. refusal to pay its dues to the United Nations, Atwood said, was "unconscionable and outrageous."

And yet economists boast that the world economy has gone up

eight times and world trade twenty times between 1950 and 2000—but to what end? The main beneficiaries of this unmitigated growth are the already wealthy, as the rich get richer and the poor get poorer. A 1999 *Forbes* magazine survey showed that the wealthiest 225 individuals in the world, mostly Americans, have a net worth of one trillion dollars; this is greater than the annual earnings of half the world—nearly three billion people!

William Vogt correctly perceived most of these trends back in 1948, only things then were not as drastic as they are now.   More than ever, we need to place our real needs before economic self-interest. Not to do so will almost certainly lead to ever greater human and ecological violence which nobody in his right mind would want to be a part of. We have no choice but to step out of this self-created *Titanic* and to search for the lifeboats that could give us a new lease on life. Rearranging the deck chairs and playing the music block the actions we need to take to reconstitute our civilization sensibly, so that we may re-inherit the Earth. We may need nothing less than a global green republic for solutions.

The Millennial Snapshot

Imagine you are an interstellar or interdimensional alien who regularly visits our home planet at the turn of each century. As always, you see this fertile, awesomely beautiful globe inviting you to have a look. Assuming you have the technological stealth to evade various radars, Star Wars defense systems and smart bombs, you begin to fly over the land to size up the situation. In some areas you see nature abound and people and animals in balance with it. In others, you see incredible structures and bustling beehive type activity dominated by industrious humans, in places interspersed with groomed patches of prosperity. And in some locations, especially in the Third World, you view filthiness, squalor, haze, ugliness and pollution. You begin to wonder, how did it all get this way so suddenly and where is this civilization headed?

You notice that the carbon dioxide levels are going up dramatically, following the examples of Venus and Mars. Both these companions to our world have atmospheric evolutions toward almost pure carbon dioxide atmospheres. Both planets are devoid of an ozone layer which would protect life from destructive ultraviolet radiation. Is the Earth going down a similar path? For whatever reason, Venus has outgassed so much carbon dioxide that it now has a hundred Earth atmospheres of the stuff, still 300,000 times more than what we have here. Because of its own greenhouse effect, the temperature at the Venusian surface is a scorching thousand degrees Fahrenheit, hot enough to melt lead.

Planetary scientists on Earth have already deduced that Venus experienced a runaway greenhouse effect. As more and more carbon dioxide was pumped into the Venusian atmosphere, the more infrared radiation became trapped, causing the temperature to rise further, leading to even more outgassing and so forth. How all this happened was a mystery to the scientists on Earth. Could it be that humans are unaware they are beginning that very same process on their own planet?

You also see Mars, once an abode for life with its abundant liquid water lakes, streams and oceans, along with an atmosphere several times thicker. Now Mars is a parched world with an evaporated atmosphere. Where had all the water gone? No liquid water is left, only the occasional winter and polar ice patch, along with a trace left as vapor. Carbon dioxide is the only gas that is left in any significant amount in the Martian atmosphere, now over a hundred times thinner than Earth's. Do the Earthbound scientists know, or do they want to know, what really happened to Mars? [19]

Some scientists on Earth talk about terraforming Mars so that it can become more Earthlike. Terraforming is an atmospheric engineering project that might make it possible for life as we know it to live on the Martian surface. For example, spreading lampblack over the polar caps would absorb more solar radiation to heat up the surface so that the carbon dioxide and water sublime into the atmosphere, creating a greenhouse effect and a more temperate climate supportive of liquid water, oxygen and life.

But what Earthlings really need to do is not to terraform Mars to be more Earthlike. They need to terraform the Earth so it can become more Earthlike—the way it used to be. In fact, Earth is becoming more Marslike. The amount of carbon dioxide here has gone up twenty per cent during this past century. Marslike deserts are growing on Earth. Some scientists predict that the breadbasket of America will probably become a dustbowl within the next decade. Could the sister planets be giving humans clues about the future of the Earth?

On your previous visit to Earth you had a very different experience. A century is a mere hairline crack in this planet's natural history. As you begin to take note of what's new, you are in for a big shock. Never have the changes been so great. In 1900, there was no Star Wars project, no radar, no airplanes in the skies and few machines on the ground. The beehives were distinctly smaller, slower and quieter, with occasional puffs of smoke coming from pockets of poverty and the cradles of a spreading industrial revolution. You heard the clop clop of horses. Nature was untouched in many more places, the population was four times lower.

Had you visited earlier, the differences would have been even greater. Is this millennial snapshot some sort of Great Experiment or nightmare or what else, you ask? The more you look at the global situation in 2000 A.D. the more you realize that it is a nightmare, as you begin to examine the stockpiles of murderous weapons, the hunger and homelessness, the toxic and radioactive waste, the overpopulation and pollution, the climate change, the poverty alongside the prosperity, the obliviousness of how highly organized and seemingly enlightened human institutions could set up their own future destruction by ignoring the warning signs which are immediately apparent to you.

Perhaps you could land on the White House lawn and share some of these concerns with the President. Assuming you could get beyond the cultural paranoia that you might stage an Independence Day type of attack, you soon discover that the President's agenda is already filled: he's too busy fighting partisan squabbles, meeting with industrial lobbyists, attending fundraisers, strategizing wars in remote oil fiefdoms and propping up the wish for a robust consumer economy. The leaders on Earth posture to be economically powerful and politically correct. They have no time to examine the deeper dynamics of what is really going on, no time to discuss or debate solutions which might eventually unseat vested interests that underlie the political power in the first place, no gumption to try on new ideas, no acknowledgement that you and your visits might even exist, outside of their heavily protected and elite cult of intelligence operatives. This has become a crazy place, you think.

Worldwatch's Gary Gardner and Payal Sampat put the situation this way: "Given the record of this century, an extraterrestrial observer might conclude that the conversion of raw materials to wastes—often toxic ones—is the real purpose of human economic activity."[20]

So you discreetly go to Congress, to leading corporate, media, financial and university suites, to 10 Downing Street, the Kremlin and other world capitals to see if you can make any more sense of the situation. You wonder how an apparently intelligent and resourceful species has organized itself sufficiently to take over a planet and destroy it. You desperately search for light in the darkness, a flickering hope that could save the day. You see it, but it's subtle, it's suppressed by very powerful forces. You realize that the planet is plunged into a dark age and its very survival is in question. Your next visit may either reveal a shattered ruin of an experiment gone awry, another return to an inhospitable Venus or Mars...or perhaps a return to Eden, the dawning of a new day, brighter by contrast because Earthlings may somehow still be able to turn away from the darkness.

You think, it almost certainly has to be one or the other, depending on what humans decide to do. They have no idea how precious their home really is.

Where are the Environmentalists?

During 1975 I was an energy advisor and speechwriter for the late Morris Udall, who was a presidential candidate and chairman of the U.S. House Interior Committee's Subcommittee on Energy and the Environment. At that time there was a perception that we had an energy crisis. The OPEC oil cartel raised their prices, gasoline shortages caused long lines at the petrol pump, and people wanted answers. This was when I received a firsthand education on the energy-environment crisis. Both fossil fuel and nuclear power options were shown to be very destructive approaches in the long run, and some of the Congressmen were looking for renewable energy options and tougher environmental regulations.

A growing and robust environmental movement was beginning to make a difference in Washington. But then something happened. By the time Ronald Reagan was elected U.S. President in 1980 and ever since, the entire environmental cause has slipped into ineffectiveness. To be sure, the Greens have won battles along the way, but the overall situation only keeps getting worse. Administrations haven't been enforcing the laws and legal loopholes appear everywhere in newer legislation. Mark Dowie well documents this situation in his book *Losing Ground*: "The Reagan years, I postulate, were the decisive decade for the mainstream environmental movement. Instead of going mano a mano with the most environmentally hostile president in recent history, the movement blinked."[21] (p.7)

"Unlike the other new social movements of the 1960s and 1970s (women's, peace, civil rights and gay liberation)" Dowie said, "which are essentially radical, the ecology movement was saddled from the start with conservative traditions formed by a bipartisan, mostly white, male leadership...rarely have they challenged the fundamental canons of western civilization or the economic orthodoxy of welfare capitalism (merchantilism)—the ecologically destructive system that gives the nation's resources away to any corporation with the desire and technology to develop them." [21] (p.28).

According to Dowie, the traditional environmental movement has mostly become a gentlemen's club of lawyers and businessmen inside the Washington Beltway, where decisions are made in boardrooms rather than courtrooms, and "compromise" is the name of the game. The statistics speak for themselves: the environmental movement has proven to be no match for what the polluters are doing. I've also noticed that most environmentalists have not looked at promising approaches such as clean ("free") energy, cold fusion, hydrogen technologies and hemp, described later in this book. They lack vision, scientific understanding, and most important, the courage to stand up for what's needed. Dowie sees some hope coming from emerging grassroots local activists who are challenging, often in court, toxic dumpers that threaten their own public health. Perhaps this movement can expand to address the needed fundamental change in our global environmental practices and the solutions. Meanwhile we continue to flounder in our own regressive thinking and practices.

I hope you will agree that it doesn't have to be that way. Waiting in the wings for their cultural opportunity, are ways out of our planetary paralysis, poverty and pollution. It is inevitable that we will be rolling up our sleeves and addressing the hard questions of the relationship we have with each other and with the Earth. How can we find more effective agendas for the future? How can we join in peace and re-direct our war machines and polluting habits to re-creating a sustainable and beautiful environment?

I believe we have an impressive range of promising approaches, on which a consensus may build. The situation calls for a wholistic viewpoint which blends technology with human development and social invention on a global and local basis. The new perspective should combine common sense about awareness of the problem, with an openness to examining solutions, including new ideas.

The Mandate

We in the West are more prosperous than any civilization in recorded history. As never before, we now have an opportunity to put that prosperity to use, for our human potential is limitless, far more than any of us can imagine. Just as humans travelled to the Moon, so we can together raise our consciousness, to become free, to enjoy the fruits of our being in concert with nature. And we can re-inherit the Earth. But before that can happen, we will need to re-qualify ourselves as passengers upon our home planet. It is time for us to give back to the Earth.

And it is time for us to grow up about that. Planet Earth, and therefore we, are in a perilous state unless our species becomes aware of how it is ravaging her resources and her beauty. In our obsession about economics, petroleum is sometimes cheaper than Perrier water. We Westerners are lulled into ignoring a basic millennial truth: that our environment is being destroyed. Instead of addressing that question in the depth that it deserves, we have made consumerism our god as it spreads its addictive candy throughout the world.

The next chapters will give us hope that we do have the technologies to restore the Earth with sustainable development in energy, agriculture, water, forestry, crops, fisheries, mineral resources, while conserving the wilderness and diversity of species. New energy technologies, hemp cultivation, biospheric restoration and opening ourselves to an emerging science of consciousness will help us enhance our our ability to came back into balance with the Earth and with ourselves. In Chapter 5 we will look at ways in which we as individuals can become healthier, happier and more aware of who we really are and where our responsibilities lie. These steps will then enable us to come together to implement solutions under new governmental and industrial structures (Part II). As in any time of great change in history, those structures will need to be built from the bottom up, and come from a motivated civil sector.

References for Chapter 1

1. Lester R. Brown et al, *State of the World* 1999, Worldwatch Institute, Norton, New York, 1999.

2. Rachel Carson, *Silent Spring*, Houghton Mifflin, Boston, MA, 1964.

3. William Vogt, *Road to Survival*, Sloane, New York, 1948.

4. Christoper Flavin and Seth Dunn, "Reinventing the Energy System", *State of the World 1999*, Worldwatch Institute, Norton, New York, 1999.

5. Ross Gelbspan, *The Heat is On: The Climate Crisis, The Cover-up, The Prescription*, Perseus Books, Reading, MA, 1998.

6. H. Josef Hebert, "Global Warming Theory Affirmed: Scientific Panel Increases Projection of Rising Temperatures", *Associated Press*, Washington, October 26, 2000.

7. Peter Bunyard, "How Climate Change Could Spiral out of Control, " *The Ecologist*, vol. 29, no. 2, 1999.

8. Peter Bunyard, "How Ozone Depletion Increases Global Warming, *ibid*.

9. "Study Sheds Light on Global Warming", *Reuters*, London, November 17, 2000.

10. "Experts: Global Warming Now Critical, Action Needed", *Reuters*, London, December 23, 1999.

11. S. Fred Singer, *Hot Talk, Cold Science*, The Independent Institute, San Francisco, 1998; Simon Retallak, "How US Politics is Letting the World Down", *The Ecologist*, vol. 29, no. 2, 1999.

12. Brian O'Leary, *Mars 1999*, Stackpole Books, Harrisburg, Pennsylvania, 1987.

13. Eugene Linden, *The Future in Plain Sight: Nine Clues to the Coming Instability*, Simon & Schuster, New York, 1998.

14. Peter Russell, *The White Hole in Time*, Harper, San Francisco, 1992.

15. Terence McKenna, "Time", chapter 8 in *The Evolutionary Mind*, Trialogue Press, Santa Cruz, California, 1998.

16. Donnella H. Meadows, "No Point Waiting Around for Leadership," *The Global Citizen*, Hartland Four Corners, Vermont, November 30, 2000.

17. "Groups Criticize U.K. Renewables Strategy", Reuters, London, November 22, 2000.

18. Zac Goldsmith, www.theecologist.org, November, 2000.

19. John E. Brandenberg, Monica Rix Paxson and Steve Corrick (editor), *Dead Mars, Dying Earth*

20. Gary Gardner and Payal Sampat, "Forging a Sustainable Materials Economy", chapter 3 in *State of the World 1999*, Norton, New York, 1999.

21. Mark Dowie, *Losing Ground: American Environmentalism at the Close of the Twentieth Century*, MIT Press, Cambridge, Massachusetts, 1996.

2

Develop Non-polluting Energy

"The resistance to a new idea increases as the square of its importance."

-Bertrand Russell

"Man has lost the capacity to foresee and forestall. He will end up destroying the earth."

-Albert Schweitzer

AS A NASA astronaut appointee in the Apollo program in 1967, I felt proud to be on a team with a positive and focussed vision: to land a man on the Moon and return him to Earth safely by the end of the decade. In those days I felt a sense of technological optimism, that in our enterprising wisdom, if a problem came up, we would find a solution. We in fact got there ahead of schedule and within budget. It all was a great success.

I generalized my positiveness to the free enterprise culture. For example, if a given approach to energy production were bad for our health, we would explore cleaner alternatives that would eventually compete in the marketplace. The government would give a special boost to those research efforts and those companies that could solve the problem. We all would want to fully explore the most promising directions— ones that would help both the economy and the ecology.

I carried my idealism into the 1970s while working on Udall's committee to develop renewable alternatives to fossil fuels and nuclear power. But, as we have seen, little has been done to bring clean and renewable options to the marketplace, in spite of enormous efforts to do so and with little Government support. For all practical purposes, fossil fuels still represent the only show in town. I slowly learned that the free market may not be so free when it comes to the multi-trillion dollar vested interests.

How wrong I was to have been so optimistic...and how right I could be, given a positive cultural political or economic shift in attitude, policy, new ideas and creative collaboration between the public and private sectors.

The world energy establishment has become enormous and unassailable, above public discussion. In 1997 seven of the twelve largest corporations in the world provided either fossil fuels or automobiles and four others are involved with related financial infrastructures. [1] Through elaborate public relations campaigns and influencing politicians and media, the energy monopoly wants you to think that there is only one way: central station power plants and distributed internal combustion engines, both burning dirty and unrenewable fuels. Nothing could be further from the truth. Renewable and clean energy is feasible and cost-effective. Even newer sources now being researched promise more elegant solutions. This chapter will show that the fossil fuel binge could come to an end, soon to be replaced with sustainable energy. This is an idea whose time has come.

The New Energy Crisis

History teaches us that the bigger an institution is and the longer it has been around, the more implacable, unwieldy and corrupt it becomes. The vested interests become a tyranny, an ego trip for those on the top. "Power corrupts and absolute power corrupts absolutely." It is therefore not surprising that, once we awaken to solutions, the very success of institutionalized "profitable" extractive processes may also become their downfall. Deep inside we all can sense that the end of the pollution paradigm may be near.

A poignant example of an awakening public was the 2001 California energy crisis, which had led to drastic price increases for an infuriated, impoverished and blacked-out public. Once one cuts through the greed and propaganda, it becomes obvious that the only solution that makes sense is a shift to clean renewables.

It follows that the wealthiest entities in the world, the energy companies, the U.S. Government and their media mouthpieces are not where you're going to find the answers. There's too much at stake for these multitrillion dollar giants to want to give up their power. Whether it's Exxon, General Motors, the U.S. Department of Energy, the President, the Prime Minister, Congress or Parliament, we can expect to get the stone wall of suppression of viable alternatives to the internal combustion engine and to the fossil-fuel-fired power plant.

These energy developments come from nineteenth century science. Fossilized hydrocarbon fuel is ignited to either: (1) explode to move some parts to run a vehicle, or (2) boil water to turn turbines which generate and pipe electricity through a grid system. This was the choice we made about 100 years ago that would to a large extent define a whole century. Perhaps the die was cast when U.S. industrialist J.P. Morgan withdrew Nicola Tesla's funding because of Tesla's interest in researching wireless "free" energy. Morgan owned most of the copper mines that would be used profitably as wire in the rapidly growing electrical grid system, made possible ironically by Telsa's own invention of alternating current for grids in 1882.

Since Tesla's time, there have been ample opportunities to look at new ways of providing energy, but most all of them have been beaten down. My own research confirms over and over that the suppression of clean alternatives is more robust than ever. Even our efforts to reduce emissions and improve energy efficiency have met with great resistance economically and politically. But our awakening to the truth will prevail as the facts become more obvious to more people. In Chapter 4, we'll return to the question of the shift in economic and political power needed to move on to a sustainable future.

Along with many other analysts of our energy-environment picture, I believe that merely reducing emissions or improving efficiency will be too little too late. No number of fuel-efficient automobiles or powerplants, catalytic converters, stack scrubbers, batteries or fuel cells could replace the deployment of clean, renewable, inexpensive energy sources. We need both. Dennis Weaver, actor and founding president of the Institute of Ecolonomics, put the situation this way: "We've tried the band-aid approach for solving the smog problem for years and predictably it's not working....(efficiency) can no longer be our thrust. We need new technology for a New Millennium."[2]

What Then are the Solutions?

No true energy expert would deny that we can create a clean and renewable energy supply which can be ultimately cost-effective, safe, and can allow us to take the single largest step to create a sustainable future. This has been known for a long time, and now the options and competitive opportunities are increasing dramatically. They fall into three basic categories: renewable energy, the hydrogen economy and new energy.

Currently available renewable energy.

The feasible renewable alternatives which have been studied and used for decades include hydroelectric power, solar energy, wind power, tidal power, geothermal power, ocean thermal gradients and solar power satellites which could beam microwaves to the Earth's surface. Each of these sources has its advantages and disadvantages. While conventional renewable energy options are basically clean and sustainable, they are often susceptible to high capital costs, diffusiveness and intermittence. They can also appreciably alter the landscape with unsightly dams, turbines, windmills and solar farms. Our choices among the renewables will require a careful look at each option and its environmental impact. Meanwhile, government subsidies and market manipulation have kept the fossil fuel industry in power, and the 18 per cent we use for renewables globally is not increasing. Only with a public awakening can these numbers change.

The hydrogen economy.

This is the middle ground. As I describe in a later section, hydrogen is clean-burning, abundant, feasible, and potentially economical. This light gas could soon replace fossil fuels as our primary energy carrier for internal combustion, power plant boilers and fuel cells. The scientific community knows the methods of hydrogen production, infrastructure, storage and consumption. The concepts involve basic chemistry and there are no show stoppers, except for the will to invest in new engineering. Commercial prototype hydrogen automobiles are commonplace. A hydrogen economy can also lead to a new energy economy through the use of cold fusion cells and hydrogen cells which could provide an abundance of cheap hydrogen and electricity.

New Energy.

Ultimately we shall be moving beyond conventional energy generation into new energy. Some of us are using the definition of new energy as "a source of energy of practical use that has heretofore been unrecognized by science." [3] By this definition conventional renewable energy is not new energy. Included under new energy are cold fusion technology, plasma or gas cells, charge clusters, underwater arcing, magnetic motors, solid state devices, electrostatic devices, hydrosonic devices and many others (see Appendix I for details about the state of the art). Each of these approaches appears to produce much more energy than can be accounted for by traditional physics, although a number of viable theories are being published in the peer-reviewed literature. Nearly every option has been suppressed, underfunded and excluded from public discussion, the media and the classroom. [4]

People often wonder why we don't have new energy if it's real and economically viable in the long run. This is because there has been little funding for the research. We are in the research phase of a development cycle that requires some investment, so big moneyed interests would rather continue with a sure thing: the fossil fuels. What is "credible" is too often defined by skeptical scientists, the vested polluting elite, and their entrenched environmentalist critics who distrust new technologies. Meanwhile, the loose assemblage of new energy inventors, researchers, communicators and educators is itself divided. Each has become his own advocate, cut off from the teamwork and the synergy that it really takes. As a result, each separates himself from the vision about the best mix of approaches. Each has desperately competed for his share of a very limited funding pie. Each has tended to be too optimistic about transforming a successful research device into a commercial prototype. Most funders become disenchanted about no return on their investment and are perplexed about the esoteric and often elusive scientific principles of new energy technologies. New energy is a difficult subject. Even with all my physics and energy background I have not yet fully comprehended the many theories, experiments and promising technologies.

Moreover, any initiative to educate the public loses its credibility and punch because we don't have commercial devices in operation, and the experimental results have been excluded from the most influential mainstream professional literature, although that too is beginning to change. Therefore the world is ignorant about all this—especially those individuals who could make the most difference—the scientists themselves and their funders. "Potential new investigators at universities and in companies have no way of developing an independent judgment of our field", says Eugene Mallove, editor of *Infinite Energy Magazine*. "Equally serious, those who might financially support cold fusion/new energy must climb a very steep learning curve before even thinking about investing or rendering aid. Few survive the high altitude ascent." [3]

New energy could be the cornerstone of sustainable solutions for a troubled world. The complexity of the technologies cannot be as easily explained as that of solar cells, windmills, biomass and hydrogen generators, gas turbines or nuclear reactors now in operation. These mechanical, chemical and physical systems are manageably acceptable to scientists and to the public. New energy escapes such familiarity and often breeds contempt and apathy. New energy experiments can produce surprising anomalies that transcend existing theory. The mainstream media won't even touch the subject. Many of us are excluded by universities from talking about it with colleagues. I often have difficulty presenting the technical essence of new energy concepts without the fear of putting my audience to sleep. For all these reasons I am reluctant to describe the details of all these technologies here and so refer the reader to the comprehensive review in Appendix I. However, I will give some case studies of successful new energy efforts.

Nowhere are the resistance to and promise of a new energy technology more dramatically revealed than those of the case of cold fusion. This well-researched approach has the potential of reversing much of the pollution while turning the interests of the energy monopolies upside down. Unfortunately, even the environmentalists haven't yet given new energy alternatives a fair look.

The Cold Fusion Revolution

The unfolding cold fusion saga has provided us with an illustrious thirteen year history that would make the suppression of Tesla seem like a school exercise. The censorship of cold fusion reveals a resistance so strong that Sir Arthur C. Clarke, author of *2001*, calls it "one of the greatest scandals in the history of science."[5]

It all began in March 1989, when two University of Utah chemists Martin Fleischmann and Stanley Pons announced a successful experiment in which a solution containing heavy water, when placed in contact with a palladium alloy cathode, occasionally produced significantly more heat energy than could have come from purely chemical reactions. They also found neutrons and tritium, telltale signs that more-powerful nuclear reactions had occurred. Only insignificant traces of radioactivity were found, contrary to what one would expect in a "hot" fusion or fission reaction. This all turned out to be both an extraordinary breakthrough for new energy enthusiasts and a scientific puzzle that confounded the theorists. Ironically, the massive Exxon Valdez oil spill in Alaska happened the very next day.

The announcement shook the world in its implications. Had we found a simple way to overcome the nuclear "Coulomb barrier" which had prevented nuclear reactions from occurring at any but the hottest temperatures, as in a hydrogen bomb or center of the Sun? It was clear that if this experiment could be substantiated, we would have a source of self-perpetuating energy whose potential is so vast, this could spell the end of energy pollution. Observers were amazed that it could happen within a cell at room temperature on top of a laboratory table!

For over fifty years, a scientific establishment had been formed around a more familiar form of nuclear energy. During the Manhattan Project of the 1940s, a tiger team of scientists developed a fission bomb by splitting atoms of uranium and plutonium. Later they came up with a fusion bomb by explosively transforming hydrogen to helium atoms, with the release of even greater energy. The former was dropped on Japan twice with devastating consequences. The latter has been deployed for potential use in World War III. There's enough of this kind of firepower to kill us all many times over, a madness called Mutual Assured Destruction (MAD).

Scientists were also able to control the fission process by placing rods of fissionable materials together just enough to cause a chain reaction which boils water in an otherwise conventional power plant, thus producing electricity. Nuclear reactors provide 6 per cent of the world's electricity (74 per cent comes from fossil fuels, and 18 per cent is renewable, mostly hydroelectric). While often on an economic par with fossil fuels, nuclear reactors have problems of their own: reactor safety (we all remember the Chernobyl disaster), radioactive waste disposal, and the international proliferation of the technology that could be used for bombs. None of these problems is trivial, which is the main reason why the Government-blessed nuclear industry, as powerful as it is, has never caught up with fossil fuels for generating electricity.

Meanwhile, some nuclear scientists have built careers trying to create a controlled fusion reaction to generate electricity. If feasible, this approach would have only some of the problems of fission. Funded at over a billion dollars a year for two decades, their attempts to confine a hot hydrogen gas for long enough to trigger nuclear reactions have met with total disappointment. Energy "breakeven", in which the reactions sustain themselves, remains elusive to this day. This has caused much frustration of those so long vested in the success of this effort. Funding is dropping and prospects look gloomy.

So along come these two upstarts from Utah who claim to be able to trigger these same nuclear reactions from heavy water at room temperature at thousands of times less cost and with no radioactivity—too good to be true! What's even more outrageous is that the claims of Fleischmann and Pons violated the known laws of nuclear physics. Looked at from the point of view of the hot fusion physicists, the Utah claims were totally off-the-wall and could be easily disproved by attempts to replicate the experiment.

Given this attitude, you can probably guess what happened next. "The claim of a chemically-assisted nuclear fusion reaction with net energy release," wrote Eugene Mallove, "threatened to divert Congressional funding from the hot fusion program. With private zeal, and later public scorn, scientists supported by the hot fusion program—particularly at MIT—sought errors in the Fleischmann-Pons work." [6]

Within a month the hot fusioneers claimed not to have been able to replicate the original results, citing the original experiment as a "possible fraud", "scam" and "scientific schlock". The U.S. Department of Energy (DOE) convened a panel comprised mostly of hot fusion scientists, who seemed to nail the lid on the coffin of cold fusion. According to Mallove, the negative DOE report had the following consequences: 1) No special funding by the U.S. Government for further research; 2) Flat denial by the U.S. Patent Office of any application mentioning cold fusion; 3) Suppression of research on the phenomenon in government laboratories; 4) Citation of cold fusion as "pathological science" or "fraud" in numerous books and articles critical of cold fusion in general, and of Fleischmann and Pons in particular."[6]

One of the DOE panel members, Prof. Steven Koonin of Caltech (and now Provost there), said, "My conclusion is that the experiments are just wrong and that we are suffering from the incompetence and delusion of Doctors Pons and Fleischmann...it's all very well to theorize how cold fusion in a palladium cathode might take place...one could also theorize about how pigs would behave if they had wings. But pigs don't have wings."[7] As a result of this scientific whitewash, for all practical purposes, cold fusion lost all credibility with the media, politicians, mainstream scientists, environmentalists and most of the public.

But throughout all the turmoil, other things were beginning to happen. Several positive observations of the Fleischmann-Pons Effect were coming from places as disparate as the Stanford Research Institute, the Los Alamos and Oak Ridge National Laboratories, the University of Illinois, Texas A&M University, the U.S. Naval Weapons Research Laboratory, the French Atomic Agency and Hokkaido University in Japan. With up to twelve years now under their belts, hundreds of scientists worldwide were beginning to discover successful low energy nuclear reactions and cold fusion in their own laboratories with a dizzying array of approaches and effects which, though they break the theoretical rules of the nuclear establishment, have nonetheless proven to work. Spinoff technologies such as the remediation of high-level radioactive waste are beginning to be patented and developed.

Respected British new energy researcher Harold Aspden, formerly IBM's European Patent Director, recently put the cold fusion flap this way: "The hot fusion community was beside itself, outraged at the audacity of such a prospect. There was a conflict of interest, tempered by disbelief, and so we have witnessed a chapter in science that is quite shameful, besides being severely detrimental to our quest to find new non-polluting sources of energy...A drastic shake-up is needed to get energy science back on course."[8]

Hydrogen Gas Cells: The Trump Card Technology?

In spite of all our confusion about cold fusion, one technology that may soon break into the marketplace involves a novel manipulation of hydrogen which can produce significant amounts of energy. When heated in the presence of a common chemical compound, the hydrogen could release hundreds to thousands of times more energy than burning ordinary hydrogen. Efforts to manufacture and sell these hydrogen gas cells to produce electricity, lighting and other products appear to be promising.

This new concept is the brainchild of Dr. Randell Mills [9], whose theoretical and experimental work have somehow survived the suppression syndrome that has befallen all new energy geniuses since the time of Tesla. In 1996, Mills founded BlackLight Power, Incorporated, called such because of the large ultraviolet emissions from his hydrogen gas and plasma. The company opened its new headquarters near Princeton, New Jersey in 1999. I often wonder what my former colleagues in the Princeton University physics department think of Mills' radical technology, how they might feel when they drive by his 53,000 square foot building, or how they might interact with any of his full time staff of thirty-five people. Regardless of what my fellow physicists might think, the Mills team truly represents the first new energy effort that has gotten off the ground.

Mills' theory behind this unexpected discovery is controversial yet plausible. He hypothesizes that, under the right chemical and thermal conditions, the hydrogen atom itself shrinks to a lower energy state with the release of large amounts of energy—yet with none of the many problems of a hot fusion reactor such as the Tokamak down the street from Mills. The process is clean and the only byproduct seems to be an inert form of collapsed hydrogen which he calls a hydrino.

At first blush, this sounds like science fiction. Regardless of whether or not the theory is confirmed, the experiments clearly show excess energy. As in the case of a cold fusion cell or ordinary fuel cell, the researchers produce hydrogen by sending an electric current into water (electrolysis). The released hydrogen from the water can then be heated and combined with a catalyst to manipulate the atoms to shrink and release the required energy. The energy output is greater than that needed to separate and heat up the hydrogen in the first place. The BlackLight Power team has shown that the energy content of water as a fuel is several hundred to several thousand times that of crude oil. One cup of water could heat and provide electricity for an entire home for one month! All this could spell the end of the fossil fuel age, in spite of the initial disbelief at first, common to any bold new discovery in science and technology.

At the time of this writing, the Mills cell is not quite commercially available but it is close. If successful, these efforts could trump every unrenewable and renewable energy option which has been scrutinized, advocated and developed. All other energy technologies would look crude by comparison and could be abandoned once and for all. This discovery would then become as historically significant as the invention of the steam engine, internal combustion engine, electric light, alternating current, solar photovoltaics, transistors, lasers, computers, jet engines, and fuel cells. If produced in the billions of units, the unorthodox Mills cells could provide abundant, clean and cheap electricity to suit whatever scale of end use is needed, whether it be used in cars, buildings or utilities. I give more details in Appendix I.

Other New Energy Approaches

The list of cold fusion, hydrogen cells and other new energy approaches and how they work is too technical and too extensive to probe deeply in this book. While cold fusion involves the principle of chemically assisted nuclear reactions in water and while hydrogen can be manipulated within water to produce excess energy, other new energy technologies appear to show energy coming from the vacuum of space. I describe the wide variety of approaches in Appendix I.

The theory in most cases is admittedly undeveloped. The most popular one posits that all of time and space contains an enormous potential energy field, virtually unknown because it behaves utterly the same in all places and in all directions. But if one accelerates electromagnetic charges , such as in plasma discharges, solid state electron oscillations and the rotary motion of magnetic motors, there appear to be circumstances under which we can tap that potential energy from the vacuum—something from what we thought was nothing! [4] As we'll see, this invisible "zero point field" (ZPF) might actually be seething with energy, a principle deduced from quantum physics.

Many vacuum energy researchers are working with electromagnetic solid state devices that apparently resonate with the ZPF with the attendant release of a lot of energy. Longtime new energy pioneer Dr. Thomas Bearden [10] and his collaborators in Huntsville, Alabama, have built a device that uses a permanent magnet which interacts with varying electrical currents that run through a coil surrounding a special material. The result is abundant and clean energy. Once the machine is started from a battery it becomes self-running. If mass-produced, this device could provide cost-effective power in vehicles, buildings and utilities.

The Bearden device could even trump the Mills technology. In principle, it could be easily scaled up to any use and has no moving parts. There would be no need to handle hydrogen or heat up anything. Imagine the day when we develop power packs that could run anything, replacing the enormously awkward polluting systems borrowed from the nineteenth century: the internal combustion engine, the steam power plant and the ugly grid system. Science fiction? Perhaps, but becoming more fact with every discovery.

The novel hydrogen chemistry and/or cold fusion technologies might also come from the zero point field itself or involve atomic transitions to so-called ground states of energy which are even lower than the ones currently known—the Mills hydrino theory. We'll return to some of these questions in Chapter 6.

It is clear that the new energy field is full of activity in spite of all the naysaying among mainstream scientists and continuing media blackouts. Here we see a complex array of technologies emerging from the research, only one of which may be all that is necessary to send us down a path of clean energy. Cold fusion and novel hydrogen chemistry may be furthest along scientifically, but my hopes lie more in solid state approaches that would involve only electricity and magnetism and no chemical or nuclear reactions. Some day we may be replacing our circuit breakers, internal combustion engines and batteries with small power units that deliver the needed electricity, in much the same way as a solar photovoltaic cell. Except this time we won't need the Sun.

For up-to-date developments especially on cold fusion and its coverups, I recommend *Infinite Energy* magazine. My previous book *Miracle in the Void* [4] describes my visits to new energy researchers on five continents. The review article in Appendix I summarizes the state of the art in 1999. I also recommend *The Coming Energy Revolution* by journalist Jeane Manning [11] who interviewed over thirty new energy pioneers worldwide. The essence of this field is that, in spite of what you might hear or don't hear from the mainstream, the technology is alive but "on life support" as one scientist puts it—perhaps with the exception of BlackLight Power. Because the research is scattered and largely unsupported, and hobbling along with a prematurely besmirched reputation, we all await and would welcome a serious government funder or altruistic investor to get together the needed tiger team Manhattan or Apollo style.

After five years of close contact with this issue I am convinced that many Wright Brothers analogues have already flown on the new energy issue; we're just not delivering the mail and passengers yet. The journalist who covered the original Wright Brothers flight was fired from his position, another indication of why the media avoid new energy research. It took several years before mainstream scientists, politicians and media could accept the fact that aviation had great potential for humanity. We live in a time of similar breakthrough and blackout, but the stakes are much higher.

It is also a time of caution for the pioneers: first, it is easy to patent something and license it to the large companies who could produce devices in the billions. But as we have seen many times in the history of innovation, any company purchasing the technology can bury the concept. This happened when General Motors bought certain electric vehicle technologies and then suppressed them. The second caution is greed: many innovators want to win the race, make their trillion dollars and become the Bill Gates or J.P. Morgan of new energy. The stakes are too high on this to hold things back or to create new monopolies. My third caution for the inventors is to be aware of other new energy technologies and, where appropriate, to join with them in finding a solution rather than bitterly competing and keeping secrets. The best answer for society, and we hope also for the marketplace, will almost always be a blend of approaches.

We now have a technology potentially capable of supplanting our highly polluting energy infrastructure, largely ignored by virtually every powerful mainstream institution. This is not because it doesn't have intrinsic promise, but because it is perceived as a threat to the ways we currently do business. Cold fusion is but one example of what an alliance (sometimes unwitting) of powerful scientific and economic special interests can do to stop a new technology. In spite of appearances to the contrary, my years of careful research on new energy reveal an extraordinary degree of ignorance and knee-jerk negativity of its potential among mainstream scientists.

"Any sufficiently advanced technology is indistinguishable from magic" was Arthur C. Clarke's response to the disbelief. His careful look at cold fusion convinces him that it has a 99 percent chance of eventually working as an energy source. Just because the theory hasn't been sufficiently developed is no reason to reject the experiments. "Recall that the steam engine had been around for quite a while before Carnot explained exactly how it worked," wrote Clarke in a provocative essay in the mainstream magazine *Science* . [5,12]

The media have consistently ignored developments in cold fusion and new energy. A few years ago, I was interviewed by a *Washington Post* reporter about the state of the art on new energy. Though the interview lasted for three hours, not a word of it was printed. Instead he published a safe historical curiosity piece on Tesla. After three years of attempting to publish a review article on new energy in the U.S., journalist Steve Kaplan and I finally found a home for it in the United Kingdom (Ref. 13 and Appendix I).

All the environmental writers cited in the last chapter are ignorant of new energy developments. As competent as these people might be regarding issues of pollution and climate change, they seem to conform to a party line about the options. We will see that because the process of science is so fragmented, most scientists and science reporters can be ignorant of developments outside their own specialties, yet claim otherwise. They seem to presume they know something they don't know outside of their box. Scientists can be unscientific when it comes to new paradigms, and this becomes very confusing to the general public. In truth, I didn't know about the reality of new energy until I spent years of in-depth research to establish its veracity. Earlier I had been doubtful and skeptical.

The Distinction between Research, Development and Deployment

Regarding new energy, we are in the (mostly unsupported) research phase of a research and development cycle. The funds required to make the necessary advances toward development will probably be a lot less than the billions spent by the hot fusion community, but certainly a lot more than is currently allocated to scattered researchers, which adds up only to a few million dollars cumulatively, except for BlackLight Power. Most of the researchers are retired and self-funded, which means they may not be actively around for much longer.

In doing my research for *Miracle in the Void*, I quickly learned that most people have difficulty understanding the distinction between research and development. Research involves gaining a basic scientific understanding of a discovery by means of repeated experiments, exchanges with colleagues, and the eventual formulations of theories and concepts that could later lead to commercial production. Development involves the more expensive and extensive effort to develop commercial prototypes leading to practical applications of what comes from the research.

So it is too simplistic to criticize cold fusion and other new energy research solely because we don't yet have a commercial prototype for energy applications. To demand that kind of instant success would be like to challenge the Wright Brothers, soon after their maiden flights, to develop a commercial prototype, or to demand that a tinkering Edison make a commercially viable light bulb within a certain amount of time. Research and early development often require an incredible degree of tenacity and wide exploration of options before anything much visible or viable happens. But when it's time to take the concept to market, we usually need teams of scientists and engineers. So far new energy has been starved of that, and so its practitioners remain as isolated citizens, often older and living on a shoestring.

Unfortunately, many individual inventors and their investor partners are often overly optimistic about being able to move from research into development which would yield potential revenues for their work. The cash won't realistically start coming in until teams have been organized sufficiently to be competitive in a new and volatile marketplace, and even then there are no guarantees. Often, when the isolated inventor doesn't deliver his commercial prototype on time, the investor withdraws his funds. Meanwhile the horse race of proprietary competition for the Holy Grail of new energy continues. I have witnessed some contestants coming and going in search of becoming the Bill Gates of free energy, awaiting the crucial moment to enter and profit.

This really is a fantasy many misguided researchers and their supporters carry out; in the end, the only way that makes sense is to find significant research dollars of tens to hundreds of millions and to come up with several alternative concepts, with no guarantee as to which ones are most viable. Many projects are run that way, including Manhattan and Apollo, and even the Tokamak hot fusion project. Ironically, the fusion project which is still well funded is the one that isn't succeeding, whereas as the one that is blackballed is the one that is succeeding.

Meanwhile, the U.S. Government continues to fund the fossil fuel and nuclear programs to the tune of tens of billions of dollars a year. What would it take for a consensus of taxpayers to demand that the Department of Energy (or a replacement agency free of corruption) fund research on the newer options as well? Wouldn't it be a positive move for the highly capable Los Alamos National Laboratories to become a new energy research and development laboratory, while the unproductive and polluting nuclear and conventional energy work is phased out? Or should we raise funds under an emerging global green republic, as discussed in Chapter 4?

Lacking that, what seems to be required is the infusion of altruistic private funds that may lead to commercially viable products. One computer software mogul withdrew his support for new energy research when he found it wasn't far enough along to "dip into the river of optimized profits" [4]. He at first may not have realized that the financial rewards don't come in until the deployment phase. It's as if financiers are at the feeding trough waiting for people to blunder and stumble until the right system is found and is just beginning to sell like hotcakes. There is a serious void that needs filling, and someone wealthy could make a big difference by altruistically supporting the work. If you have any leads, all of humankind and the planet will thank you.

Two Groups of Scientists

So far we have looked in some detail at two scientific battles as stages on which future decisions must lie. The first were a group of mainstream climatologists who have used their special knowledge to report on an important problem, human-caused global warming and climate change, but their conclusions have been twisted and censored by powerful economic interests. I have also reported on the results of an equally qualified group of scientists who claim to have a potential answer to the energy crisis, almost entirely snuffed out by another group of scientists, the hot fusioneers and mainstream physicists and chemists, who are trying desperately to defend a diminished turf. The rest of the scientific community, media and public remain uninformed, so by default side with the skeptics.

In these and many other examples, we see that some mainstream scientists can abuse their power. They have sacrificed their quest for the truth and instead have become politicized, polarized, institutionalized, bureaucratized and bought out. Partly for that reason, the public increasingly distrusts science. Also science has led to technologies that have wreaked havoc on our environment, and so are we to trust these elitists to guide our future? What makes matters even more difficult is that many people are afraid of science, because the concepts can be complex. Yet we need to listen to climate change and new energy scientists—not the skeptics, politicians, pundits and economists who continue to calculate policies independently of what's happening to the biosphere.

As we shall see in the Chapter 6, science itself is in for an overhauling. But this doesn't negate the work of hundreds of brave climatologists and new energy researchers who have something very important to say. While the two groups don't yet interact, their clear messages now need to be put together: we have a serious problem and we also have a probable solution to that problem. That story hasn't been told and it needs to be for our own survival.

The fact is, our choices in science continue to create the circumstances of our future. Research and development form the thin edge of a wedge of what's to come in the twenty-first century. Today's million-dollar blueprints become tomorrow's multi-billion dollar projects. In the case of energy, the number goes into the trillions. Energy is by far the single largest sector of the international economy, some estimates going as high as eighty per cent of all commerce, directly or indirectly.[14] The stakes in these decisions are now about as high as at any time in human history. But we will need to know what's possible, free of suppression.

What if New Energy Options Don't Work?

We have seen that we are almost certainly on the threshold of a new energy technology that could make great strides in reversing ecocide. Here we have the potential to end most air pollution, carbon dioxide buildup, global warming, climate change, a scarred and scorched Earth, the ugly power lines, oil drills, refineries, supertankers, oil fiefdoms, refineries, gas stations, coal mines, acid rain and so on. But let's say the irrational forces of man or nature don't make that possible. Let's say, for example, that the commercial prototype or implementation are blocked by feasibility questions or by continually blind political and economic systems or by unexpected fatal flaws in all systems which would have serious ecological consequences or unstoppable weapons applications or maybe even something as bizarre as a superior alien visitor telling us to retrench. It's heartening to know that, based on the mainstream literature on energy options, we could still wiggle out of the problem, although this path would be certainly more expensive and would probably involve the more intensive use of materials.

The leading traditional renewable candidates are solar photovoltaics, wind power and biomass combustion, combined with an overhaul of the efficiency of energy systems and the use of clean hydrogen fuel. Worldwide wind power is growing by 25 per cent a year, with a $2 billion market in 1998. The use of solar photovoltaics is increasing by 17 per cent a year. While prices keep decreasing, the manufacturing cost will need to go down another 50 to 75 per cent for solar electricity to become competitive with the grid. One promising new solar technology called Lumeloid uses polarized film, which could produce up to ten times more power per unit cost than photovoltaics (www.ardev.com). Still, new energy would seem to be more attractive than either the wind or sun: the intermittent and diffuse nature of both these accepted sources require either energy storage or complementarity with each other and to the grid. Also windmills tend to consume more raw materials, litter the landscape like power lines and can pose maintenance challenges. Nevertheless, these approaches are very appealing conceptual backups until new energy comes along.

Another promising technology is fuel cells, which run on hydrogen and oxygen. The cleanly burned product is water, with the release of electricity. This process is the opposite of the electrolysis of water, where electricity dissociates the water into its constituent atoms hydrogen and oxygen. Unfortunately, we have not yet found cost-effective ways to provide the hydrogen for the fuel cells and for other energy uses. All of that could change in the fully developed hydrogen economy described in the next section. Ironically, cold fusion or Mills cell technology could provide the most inexpensive route to hydrogen production, but if we have these technologies we may not need large amounts of hydrogen for fuel anyway.

Yet another approach to traditional renewable energy is growing crops whose biomass can be burned to produce charcoal, fuel oils, process steam, methanol and various chemicals. While carbon dioxide goes into the atmosphere from burning, its equivalent amount can be reabsorbed by the new crop, so the entire process becomes greenhouse-neutral. Hemp has been shown to be a particularly efficient biomass crop.[14] In the next chapter we will be looking at many other uses of this versatile crop as a substitute for paper, clothing and construction materials.

We can also improve our energy efficiency without making any major changes in our consuming habits. Amory Lovins, director of the Rocky Mountain Institute, believes we can save at least three-quarters of our electricity use simply by redesigning power plants, buildings, appliances and heating and cooling systems, when combined with renewable energy. Lovins has also demonstrated that automobiles could be built to be lighter and more "slippery", yielding gasoline economies of several hundred miles per gallon without sacrificing comfort or cost. [15]   As an example of our energy waste, Lovins points to what happens when we create lukewarm water and air from high grade electrical energy produced in a remote power plant at thousands of degrees. "This is like using a chainsaw to cut butter," he said. [16] Improving efficiency could play a large role in shifting the energy marketplace to renewables, while saving about $300 billion a year, Lovins argues.

As of a few years ago, Lovins joined the ranks of the mainstream in opposition to new energy technologies, something I discovered one day when we happened to be on the same flight. At baggage claim, I asked him how he felt about free energy. He became very ruffled and walked away from me, even though we had known each other from before at various energy policy venues. Such behavior seems to be the party line for the environmentalists who haven't looked at the research because they have already made up their minds about our best energy future. A shift to conventional solar, wind, biomass and increased efficiency would be basically a shift from one trillion dollar industry to another, with environmental benefits. But a shift to new energy could be far cheaper and environmentally even friendlier, and more resources could be freed to restore the biosphere in other ways.

Nevertheless, it is nice to know we have a choice. Alden Meyer, an energy researcher at the Union of Concerned Scientists, estimated that a total transition to conventional renewables would only cost about $25 billion a year over the next ten years. [17] That amount is less than goes into U.S. government subsidies for the development of oil, coal and nuclear energy.[18] The implication here is that a simple shift in public policy may be all that it takes to do the job!

"The great news," wrote Ross Gelbspan, "is that each historical energy transition has resulted in an explosion of economic progress...The only remaining question is how much damage from the world's increasingly unstable climate we would have to absorb from the transition—how long it would take for the planetary atmosphere to return to stability...Scientists do not know at what point an unstable climate will become a cascade down a steep slope." [17]

Even with a rapid shift to renewables, the consensus is that it will take decades or more for the atmosphere to restabilize. While we would almost immediately notice cleaner air, most of the excess carbon dioxide could remain in the atmosphere for one to two hundred years. Replanting the forests, preserving what's there and cultivating hemp would speed up the process, but it will still take awhile. The legacy of climate change and its future uncertainty will probably be with us for a generation or more, thanks to the unwise choices we have made during the twentieth century. Our children and grandchildren will have to live with this legacy. To further delay consideration of these issues and to obfuscate them within the games giant corporations and politicians play will almost certainly lead to our own demise.

Nevertheless the more incremental approaches of Amory Lovins, the Union of Concerned Scientists and other environmentally motivated energy study groups would work in the long run. In a more mild way than the probable coming of new energy breakthroughs, their strategies could in the end help relieve us of the enormous problems of the hydrocarbon and nuclear age. But the suppressed new energy technologies will almost certainly speed the transition and make energy very cheap and convenient. We can have it either or both ways. But are we up to the task?

Some of the energy companies and governments of Europe are beginning to see the light. In 1997 British Petroleum announced it is putting $1 billion into renewable resources and Royal Dutch Shell was quick to follow with another $500 million. The German coalition of liberals and the Green party is planning a tax increase on polluting energy. The Germans are also launching a 100,000 solar rooftop program, which they expect will lead to a significant increase in the marketplace, amounting to billions of dollars. [19] Japan is also pursuing an aggressive photovoltaics program. Not bad for countries that are not exactly in the Sun Belt.

The Hydrogen Economy

We return to that simplest and most abundant element in the universe—hydrogen. It is remarkable that we have spent so much time and created so much havoc on drilling something and burning it to our peril, and all the while we have a clean fuel that is well understood and tested. Even short of embracing the strange chemistry of Randell Mills and the zero point and cold fusion results, hydrogen is the ace in the hole among the extraordinary range of clean, and renewable energy sources. It is most remarkable that the fossil fuel lobby has effectively blocked hydrogen for so long, because the technology for producing, handling and burning hydrogen is well-known. As in the case of new energy, the applications and approaches are many but the funding is sparse.

The energy establishment is not happy about hydrogen. Nejat Veziroglu, an engineering professor from the University of Miami, was nominated for a Nobel Prize in Economics for his work in pointing out that petroleum plus pollution is more expensive than implementing a clean hydrogen economy. [20]

When combined with oxygen in the air or in water, hydrogen releases energy as heat or electricity that could run cars, power plants and buildings. The technology has now progressed to the point where the only significant emission product is water vapor.

Hydrogen can come from many sources. Currently most of it is made from heating hydrocarbons, either from petroleum, coal, natural gas, and waste. It can also come from water when electricity passes through it. This is called electrolysis. The recombination of oxygen and hydrogen releases heat from combustion or electricity in a fuel cell, with the final product being water once again.

Nearly all hydrogen on Earth is combined with other elements to form compounds of very different properties—water, the hydrocarbons (including fossil fuels), natural gas and waste (methane), biomass (carbohydrates including methanol, ethanol, and ether), smelly hydrogen sulfides, pungent ammonia and stable metal hydrides. One of the current challenges is to find inexpensive ways to produce hydrogen, where the energy input is less than the output from consuming the hydrogen. Electrolysis using new energy technologies would be the most elegant approach because the energy would be cheap, clean and virtually unlimited. Short of that, the price of photovoltaics keeps going down while the price of fossil fuels keeps going up, so the costs of clean electrolysis of water and subsequent hydrogen combustion are coming into a competitive range with burning fossil fuels. Slightly dirtier is the gasification of coal (a much, much cleaner process than burning coal) and the scavenging of hydrogen from natural gas, biomass and waste.

There are many ways to store hydrogen, either as a gas, liquid or solid metal hydride. Because the hydrogen molecule is so small, hydrogen diffuses rapidly and is difficult (but not impossible, given some development) to store in tanks for long and under much pressure for everyday use such as in automobiles. Hydrogen liquids must be kept very cold and used soon because of boiloff. Both approaches could end up being as manageable as a tank of petroleum. Ignorant skeptics try to besmirch the image of hydrogen by pointing to the Hindenberg disaster. Actually, the famous fire was probably caused by a lightning strike combining with onboard diesel fuel and a highly flammable covering; it was only later that the hydrogen within the balloon caught on fire.

Nevertheless, there are safer, cheaper and more elegant ways to produce, handle and burn hydrogen than storing it in high pressure gas tanks or as a liquid. For example, hydrogen gas can be produced on demand by combining alkaline metal emulsions with water. [21] It could also come from solid metal hydrides. [22]

Perhaps no individual has had more practical experience with hydrogen than Dr. Roger Billings. [22] As a precocious child in Utah, he began experimenting with hydrogen motors in school science projects during the 1960s. He has since built over twenty hydrogen vehicles and a home hydrogen heating, cooling and cooking system. Billings tinkered with a wide variety of approaches. Through trial, error and thirty years' experience, he has come up with what he considers to be the most promising technology for automobiles: compact fuel cells with safe metal hydride storage containers to provide the hydrogen.   If mass-produced, these vehicles could provide high performance, low cost electrical operation and have a range of 300 miles before recharging the metal hydride canisters with new hydrogen from refueling stations.

We recently visited Roger Billings at his underground laboratory in Independence, Missouri. This abandoned limestone mine, about 100 feet deep, maintains a year round temperature of 58 degrees Fahrenheit. Heating costs are therefore minimal and building costs are $10 per square foot, ten times less than for most buildings today. Billings has the potential to expand the space to two million square feet. The experience reminded us of a Jules Verne novel. Who could not be excited about a real-life brilliant eccentric bearded scientist and his staff deep inside the Earth building a device to save the world, free of the pressures of our dysfunctional institutions? The lack of diurnal cycles induced in us a sense of timeless creativity and unawareness about braving the elements above. Whether it was day or night, blizzard or heat wave, tornado or hurricane, nuclear war or chemical attack, there was a comforting feeling being within the womb of Gaia. I have often thought that at the rate we're going, we may all have to live underground. Roger Billings' Hydrogen Hobbit Hollow may be another prototype of what's to come.

As in the case of Randell Mills, Billings has more potential to succeed than others because he has both money and expertise. He founded WideBand, a successful one billion bit computer networking concept he had invented. The company will soon go public and could potentially yield hundreds of millions of dollars which he would like to see used for his hydrogen work in a year or so. Like Mills, Billings has a rare blend of inventiveness, scientific sophistication and entrepreneurship.   John Bockris, Distinguished Professor of Chemistry at Texas A&M University and author of the first technical papers and books on hydrogen between 20 and 30 years ago [23], has labeled Billings as a "doer", perhaps one of the few in this complex and distracting world.

Two other American inventors worth noting are Tim Lee and Roy McAlister. Lee has built cars capable of burning various biomass products or natural gas. He then "decarbonizes" the exhaust by heating it and sieving it. [24] At this writing, Dennis Weaver plans to drive one of Lee's cars (and a pure hydrogen-burning car) on a highly publicized "Drive for Life" across the U.S.   Roy McAlister of the American Hydrogen Association has invented a "smart plug" which could ignite hydrogen in conventional internal combustion engines, and thereby eliminate NOX emissions from the burning hydrogen. [25] He can also switch over to natural gas as needed. Both the McAlister and Lee technologies could give us the potential of retrofitting up to a billion vehicles now extant on the planet.

The worldwide hydrogen community is growing rapidly.[26] Increasing fossil fuel prices, limited oil supply, decreasing fuel cell prices, prospects for low cost solar or new energy electrolysis of water plus the great abundance, safety, versatility and environmental friendliness of hydrogen make this an obvious winner in a future energy economy. For example, information is circulating on the Internet that American inventor Dan Kamen is introducing a hydrogen scooter, a development which may be supported with substantial investment. Or it is conceivable that our classical conception of a hydrogen economy by transforming ordinary hydrogen into water may be trumped by cold fusion, Randell Mills' more energetic hydrogen-to-hydrino process or other new energy initiatives such as Bearden's, based on electromagnetic interactions with the zero point field. Perhaps we are fortunate not to have developed a full fledged hydrogen economy yet. Instead, we might be able to move from fossil fuels to new energy in one elegant step. Still, I feel that a careful plan for the hydrogen economy, in parallel with supporting research and development of a new energy economy, could give us a coherent strategy to supplant fossil fuels as soon as possible. I am therefore starting a joint effort through Dennis Weaver's Institute of Ecolonomics, the International Association of New Science and others to come up with a unified and flexible future energy policy consistent with ending air pollution and global warming (Appendix II and IV).

Billings and Bearden foresee the day when hydrogen and new energy scientists can license their technologies to the large corporations that could scale up to mass production. They and others feel we must soon approach the auto, energy, aerospace and utility companies to implement clean and renewable energy technologies on a large scale as soon as possible—so long as the giant corporations don't bury the ideas. Companies already interested in the hydrogen economy include BMW, British Petroleum, Shell Hydrogen, Honda and Daimler-Chrysler. There is a lot of interest in hydrogen and solar power in Germany, particularly.

I have begun to be active in the hydrogen field because it is such an obvious interim solution to our global energy crisis. No matter what happens or doesn't happen with the more exotic new energy research, hydrogen and other renewables could fill the bill for a clean energy future. Retrofitting existing energy systems and designing new systems for transportation, utilities, and heating, cooling and cooking could all neatly unfold in a hydrogen economy which would supplant a fossil fuel/nuclear economy. New energy breakthroughs could add fuel to the fire.

While this mix of sustainable energy makes sense, their timely implementation (now is not too soon) awaits a paradigm shift in our social and political thinking. We shall see that something as simple as a populist movement to shift government subsidies from petroleum and coal ($20 billion a year in the U.S. alone) to new and renewable energy could be all that's needed to trigger the revolution (Chapter 4).

It is ironic that the greatest seat of the problem is also the greatest seat of the solution: America. We shall need to find ways to reconcile these polar opposites residing in the very same land so that we can surmount the problems and arrive at the solutions. If they cannot be reconciled here in the U.S., those of us who are solution-oriented may need to expatriate ourselves from a leadership based on destructive and unnatural principles. New energy inventors Bruce DePalma, Martin Fleischman and Stanley Pons moved offshore to carry on their work, because they had experienced great hostility and suppression here in America. I hope it doesn't have to come to that for the rest of us and that the U.S. Constitution and equal rights are upheld during the coming energy revolution.

What We Must Do Now

One way or another we have the technological wherewithal to take care of our most pressing ecological challenges by converting as soon as possible to clean and sustainable energy sources, both traditional and nontraditional. But humans must get beyond the harangues of the greenhouse skeptics and the new energy skeptics, who are retarding the research and development that are so badly needed. And we need to motivate a confused and apathetic public, particularly in the West and Japan, who will be leading the way to introduce the technologies worldwide. This must happen before China, for example, builds ever more coal power plants and purchases hundreds of millions of automobiles. The conspicuous energy consumption of America is only the start of an ugly juggernaut in bringing our global game of Russian Roulette to ever higher levels if we do not change our ways.

A scientific consensus is emerging that we must deploy clean energy technologies in a global crash program. A small price to pay might be the unravelling of those institutions vested in polluting ways. So here is the second step we must take to re-inherit the Earth:

Develop Non-polluting Energy.

For that, we must: (1) shift government subsidies from the fossil fuel/nuclear interests to new and renewable energy and a hydrogen economy, and (2) rapidly create those laboratories which can bring the new energy options closer to fruition. We

will want to see what's available now and the near future, so
we will be able to make sensible decisions, far beyond the ignorance, suppression and debunkery which have dominated the
scene until now. We must look at flexible energy infrastructures
that will allow for innovation to replace what is available at
any given time.

In some ways, this program will be easier to implement than the
more scattered local efforts and multitude of approaches implied in
renewing sustainability to agricultural lands, forests, rangelands, fisheries, waterways, wilderness areas, and resource extraction discussed in
the next chapter. Because the way we are abusing our energy worldwide
involves the same type of technology everywhere, the solutions can be
more clearly expressed and implemented globally—as they must.
Basically, we will be substituting a new set of systems for the old. The
new energy program will be the greatest and most positive undertaking
humanity has ever created. The concept of shifting subsidies and funding research and development is so utterly simple, requiring only the
actions of democratic government guided by the will of the people. Then
we can see the changes unfold rapidly.
But if our leaders insist on continuing to be bought out and the
public continues its apathy, there is an alternative: to pool some of the
great private wealth accumulated in our prosperous society into an
aggressive energy program. Perhaps only a global green republic
described in Chapter 4 could get us through the crisis. Through the sheer
power of attraction, clean and cheap energy development would put us
on the right path. Once implemented, the new program will make the
fossilized hydrocarbon industry and leading politicians look silly. The
United States sits on the crossroads of change, and has the ability to lead
the way. We shall need a blend of a freshly awakened grassroots and possible altruistic financing to move us in the right direction. The time is
now while there are still deep pockets in government and industry. The
coming instabilities, including the climate itself, may produce drastic
conditions that would preempt our ability to move beyond them to solve
the problem once and for all.
Through its new tax structures and incentives to develop solar
energy, the Germans provide an example of what must be pursued globally for our own survival. Upon this, a strong scientific consensus is
emerging, and it is beyond technology and invention. It will ultimately

depend on civil actions which will result in a restructuring of governmental, social and market systems that transcend current practices. We shall return to this question in Part II. But energy is not our only problem. Next we will look at how to redress other biospheric challenges.

References for Chapter 2

1. Christopher Flavin and Seth Dunn, "Reinventing the Energy System", Chapter 2 in *State of the World 1999*, Worldwatch Institute, Norton, New York, 1999.

2. Dennis Weaver, Institute of Ecolonomics Newsletter, May/June 1999.

3. Eugene F. Mallove, "New Energy and the News Media", *Infinite Energy*, vol. 6, issue 34, 2000.

4. Brian O'Leary, *Miracle in the Void*, Kamapuaa Press, Kihei, Hawaii, 1996.

5. Arthur C. Clarke, "Presidents, Experts, and Asteroids", *Science*, June 5, 1998.

6. Eugene F. Mallove, "Ten Years that Shook Physics", *Infinite Energy*, vol. 4,issue 24,March/April 1999, p. 3.

7. Steven Koonin, at the 1998 Baltimore American Physical Society Meeting, quoted in *Infinite Energy*, vol. 3, issue 18, 1998.

8. Harold Aspden, "Ten Years of Cold Fusion: Or Was it Ten Years of 'Cold War'?", *Infinite Energy*, vol.4, issue 24, 1999,   p,15.

9. Randell L. Mills, The Grand Unified Theory of Classical Quantum Mechanics, BlackLight Power, Inc., Cranbury, New Jersey, 2000; www.blacklightpower.com.

10. Thomas E. Bearden et al, "The Motionless Electromagnetic Generator: Extracting Energy from a Permanent Magnet with Energy-Replenishing from the Active Vacuum, "Magnetic Energy Limited, Huntsville, Alabama, 2000; also a description of my visit with Bearden is in my book *The Second Coming of Science*, North Atlantic Books, 1993.

11. Jeane Manning, *The Coming Energy Revolution*, Avery Books, New York, 1995.

12. Arthur C. Clarke, quoted in *Infinite Energy*, vol. 3, issue 20, 1998, p. 6.

13. Brian O'Leary and Stephen Kaplan, *Review of the Scientific and Medical Network*,no. 71, Colinsburgh, Scotland, Dec. 1999.

14. Jack Herrer, *The Emperor Wears No Clothes*, HEMP Publishing, Van Nuys, CA, 1990, p. 45.

15. Amory Lovins,interview with Robert Evangelista, Nevada City, CA,transcript at 1-(800) 51-SOLAR.

16. Amory Lovins, quoted in *Get a Life!* by Wayne Roberts and   Susan Brandon, Get A Life Pub., Toronto, Canada, 1995, p. 56.

17. Ross Gelbspan, *The Heat is On*, Perseus Books, Reading, MA,1998: interview with Meyer, p. 98; other quotes from pp. 103 and 181.

18. Doug Koplow, "Energy Subsidies and the Environment" in *Subsidies and the Environment*: Exploring Linkages, Organization for Economic Cooperation and Development, 1996, and quoted in reference [17], p. 98.

19. The German photovoltaic project was reported in Photovoltaic Insider's Report, December 1998.

20. T. Nejat Veziroglu, "Solar Hydrogen Energy System: A Permanent Answer toEnergy and Environment Problems", *Proceedings of the Forum on Converting to a Hydrogen Economy*, Ft. Collins, Colorado, compiledby M.Albertson and H. Jarvis, September 2000; John O'M. Bockris and T. Nejat Veziroglu with Debbi Smith, *Solar Hydrogen Energy: The Power to Save the Earth*, McDonald & Co.,London.

21. "Hydrogen Energy Innovation from Water and Air for the Twenty-First Century", Via-Tek News, vol.1, no.1, Montello,Nevada, Jan. 2001.

22. Roger E. Billings, *The Hydrogen World View*, International Academy of Science, Independence, Missouri, 2000.

23. J. O'M. Bockris, *Energy Options*, Halsted Press, New York and Taylor & Francis, London, 1980; "Hydrogen Economy in the Future", International Journal of Hydrogen Energy, no. 24, 1999.

24. Tim Lee, "Practical Approach to a Hydrogen Based Economy", Proceedings of the Forum on Converting to a Hydrogen Economy, Ft. Collins, Colorado, compiled by M. Albertson and H. Jarvis, September 2000.

25. Roy McAlister, "Hypercars", *ibid*.

26. Robert Siblerud, *Our Future is Hydrogen*, New Science Publications, Wellington, CO, 2001; Jeremy Rifkin, *The Hydrogen Economy*, Tarcher Putnam, New York, 2002.

3

Preserve, Restore and Sustain the Biosphere

"We're challenged as mankind has never been challenged before to prove our maturity and our mastery, not of nature, but of ourselves."

-Rachel Carson

AS I READ over the material of the last chapter, I often feel silenced about what our species is doing to the planet. How much more crudely do we have to act, I wonder, before we get to the point where we'll see the light and stop treating nature as something to be conquered and exploited? This idea has dominated the Western mind since Genesis, and more poignantly, since the time of Francis Bacon and Rene Descartes in the 1600s. How right these Renaissance fathers of science were about wanting to expand our knowledge about ourselves and our universe. How important was their motivation to ease our lifestyles through technology. But how utterly wrong they were to begin an era of the human abuse of the natural world to the point we are being destroyed. City dwellers today are often afraid of nature and so push it away. A paradigm shift will be necessary. We are going to have to change our attitudes about how we relate with nature and how she could relate back.

I write this from beautiful, remote Hana, Maui, Hawaii. I am irritated today with the sputterings of a utility wood chewer nearby and about fifty tourist and marijuana control helicopters over the past two hours filling the air with noxious noise pollution. Bill Ford, CEO of Ford Motor Company, predicted the demise of the internal combustion engine. [1] So why don't we get on with the job? New energy and fuel cell transportation will run on quiet electricity. It's a relief to know that most of our half billion automobiles have mufflers. Not so for helicopters and lawn equipment.

I think also of the foul air and extreme weather reported in the daily world news and can empathize with those who are not as fortunate. I feel guilty about contributing to the pollution every time I fill up my gas tank, take a jet to the mainland, turn on the oven or draw a hot bath. What it all seems to come down to is that there is a great power behind the throne suppressing the next major step for our survival. That power is a conspiracy of vested interests.

In America we at last see the tobacco lobby collapsing. Perhaps one of the next to go will be the subsidized energy lobby whose creations are causing more death and suffering than any amount of tobacco has ever done or could ever do. Living in some cities in China and India is more hazardous to the lungs than smoking a pack of cigarettes every day. Resperitory illness is the number one cause of death in these places, affecting hundreds of millions of people. We don't seem to have the time for a lengthy adversarial process in which we endlessly debate the science, deliberate about whether or not to label our gas pumps and electrical switches with warnings from the Surgeon General, and begin the slow process of gathering signatures on class-action suits which could eventually put the energy companies out of business. Or, like some of the tobacco giants, perhaps the oil giants will merely shift their cash and power elsewhere. Meanwhile the ice caps melt, the seas rise, the air is unbreathable, the water undrinkable, and the storm disasters, plagues and migrations escalate. What good is a protracted debate anyway, when dirty energy is the only show in town and the obvious solutions are being suppressed?

How do we counteract the energy lobby in its self-serving cry that the economy would be ruined were we to make the necessary shifts to renewable energy, even though the analyses of Lovins and others show just the opposite to be true? In its November 23, 1998 issue, *Time* magazine reported that leading oil, chemical and logging companies have already received over a billion dollars in tax breaks to help create new jobs. For many of these companies, the benefit becomes a great opportunity to profit, as the cost per new job to the taxpayer can run into millions of dollars! Corporate welfare. Somehow we can be more imaginative in creating jobs to restore the Earth.

We have seen that the main solution to poisoning our atmosphere is simple—implement clean energy everywhere, whether it be solar, hydrogen or an extraordinary new energy concept. We the people will need to become an educated majority at a global level, and through our elected officials, to implement the plan. We the people too are a corporation, the biggest and richest one in the world, although we don't realize it sometimes. Our inability to come to grips with making the necessary shifts seems to be our greatest downfall, a vicious cycle which suggests that as the problem escalates, the worse it will get because we become blinded with gridlock, denial and apathy. So while the solution may be simple, it somehow isn't easy. We return to that question in the next chapter.

"This gridlock", writes British author Palden Jenkins, "involves an overloading of the global system and world psyche by the sheer compromising force and intensity of events...a result of failing to foresee or to deal with problems at the time they first arise...Official decision-making bodies are not elected or constituted with a brief to make the fundamental shifts needed. They are created to conform to the specifications that were set, and no such body wishes to go out of business. Jobs could be lost—in our country! " [2]

"The natural outcome of this creeping sclerosis," Jenkins continues, "is growing structural strain of all systems, increasing degradation, disinterest and disaffection in world society, sinking vitality and suppression of fundamental innovation. This all adds up to an increasing tendency to delay action. This is what felled the Roman empire, and we're happily repeating it...People in positions of influence suddenly lack solutions: they have the sticky fingers of complicity, they do not understand the parameters of the reality which has suddenly overtaken them, and genuine truth-speakers amongst them have already been weeded out—or they resigned in disgust. Everything gets dreadfully stuck."

A *USA Today* poll reported that two-thirds of Americans believe that we will have an environmental catastrophe by 2025. Yet somehow mysteriously the same two-thirds seem to doubt we can do anything about it, even though we created it. It's as if we're all asleep at the wheel of a gigantic gas guzzling sports utility vehicle racing down a hill out of control towards the inevitable crash.

Indeed, the vitality and spirit of an Apollo-type effort appears dead, as the agendas and choices of candidates in democratic societies narrow ever more in the direction of conformity to corporate interests and outmoded structures. Our choices in politics are most often between bought-out Tweedle Dees and Tweedle Dums while the media focus on their personal affairs, minute political disagreements and spin-doctoring.

Nowhere is this more evident than in my own country, leader of the fossil fuel glut. Perhaps the only recent incumbent American politician who has taken a strong stand against hydrocarbon pollution was Al Gore. During his losing bid for the U.S. Presidency in 2000, Gore downplayed his environmentalism to the point that it was a nonissue. His political opponents, big business and the media relentlessly attacked him on his ecological bent, so he focused instead on his own political future. It is interesting to note that Gore's college thesis advisor at the University of Tennessee was Roger Revelle, one of the pioneer scientists to warn about human-induced global warming and climate change.

As an example of pack journalism, syndicated columnist George F. Will on May 2, 1999 quotes Gore from Gore's unusually strong and perceptive book *Earth in the Balance* [3]: "Automobiles pose a mortal threat to the security of every nation that is more deadly than that of any military enemy we are ever again likely to confront", to which Will responds, "Should we then be bombing Detroit instead of Serbia?" This demonstration of journalistic ignorance is the very kind of thing which could take the issue out of public debate. Perhaps acting on advice from the pundits, Gore feared the loss of votes were he to continue to speak out on the environment. Politics is always the art of the possible, not the science of what's real, and Gore might have strategized that he had the Green vote already, so why try to rock the boat? So he moved towards the perceived political center and abandoned the environment as an important issue. But his brushoff of Green Party candidate Ralph Nader may have cost him the election.

Does Mr. Gore still agree with his own statement, "We must make the rescue of the environment the central organizing principle for civilization"? *Time* magazine's Earth Day 1999 report argued that the environment should be the central focus of Gore's campaign, because it would draw fire from his opponents and bring the issue before the American people. Then the people could decide, as a true democracy should work. Or had he already received too many campaign contributions from those wealthy enough to place a silence upon him and the Green movement in America?

"I've seen Gore on TV quite a lot lately", the American environmental journalist Remy Chevalier wrote to me during the 2000 campaign. "He seems like he's on auto-pilot, or quite heavily sedated. If indeed he feels as strongly as he did in *Earth In the Balance*, he must feel beaten by all his compromises. Ah, the things you have to do to be a politician...Ask any Joe Blow on the street and he knows he's getting the shaft from the oil companies and that miracle technologies are being suppressed. Ask anyone!! They may not know much, but they know that much. So why doesn't that attitude make it up the chain of command expressed by those supposed to represent the people's will? Maybe because the people act and feel as if it's a losing battle....One problem is that it's not enough to know you're getting screwed, you have to do homework and figure out how and by whom! And that's what I propose we do, stop patronizing the public, and just give them the tools and ammunition to make a case."

Restoring Sustainability

We have seen in the last chapter that all we need to do to reverse climate change and air pollution is to shift the way in which we use energy. In this chapter we'll look at other ways to restore the vitality of the biosphere. Reclaiming biodiversity, wilderness, farmlands, grasslands, forests, aquifers, oceans, fisheries, mines and waste becomes a more complex and diverse set of tasks. Institutionally we are dealing with organizations that are more distributed and less monolithic than the energy production business, so both problem and solution are less susceptible to any simple or centralized redirection. Nevertheless, we shall find that the solutions are abundant in all areas. Once again, the discoveries of science and sensitivity to nature can assist us in terraforming the Earth to become more Earthlike, the way it used to be.

"The ecological principles of sustainability are well established, based on solid science", wrote Lester Brown and Jennifer Mitchell. [4] "Just as an aircraft must satisfy the principles of aerodynamics if it is to fly, so must an economy satisfy the principles of ecology if it is to endure. The ecological conditions that need to be satisfied are rather straightforward. Over the long term, carbon emissions cannot exceed carbon dioxide fixation; soil erosion cannot exceed new soil formed through natural processes; the harvest of forest products cannot exceed the sustainable yield of forests; the number of plant and animal species lost cannot exceed the new species formed through evolution; water pumping cannot exceed the sustainable yield of aquifers; the fish catch cannot exceed the sustainable yield of fisheries."

The Worldwatch researchers are certainly right to a first approximation. But we also need to look at more sensitive issues involving the qualities of our lives. For example, do we want to replace our precious remaining diverse primary forests with homogeneous tree farms, which the logging interests would want us to do? This might satisfy the criteria of replacing cut timber with new trees for another round of consumption and preserving a carbon dioxide sink as well, but how does it feel to walk through a densely packed monocultured pine forest planted by humans? Is there the same experience of natural diversity of species, of the quality of life? The homogeneous forests do little for biodiversity or for the naturalist in us.

Or what do we make of recycling water for agriculture through irrigation ditches and pumps and hydroelectric systems if these actions destroy the original quality of the water and land? The agricultural and water systems might be themselves sustainable, but how about the quality of the food and water or the aesthetics of the environment at the place of growth or downstream? These will be the kinds of topics we will be debating for some time. But at least Worldwatch is giving us the minimal cleanup scenario: restore the major ecological systems to sustainability.

Disturbing the natural ecology of the Earth is like peeling the layers of an onion, sometimes quite literally. For example we are ruthlessly stripping our topsoil beyond repair. And then each successive peeled layer symbolically represents an increase in the crudeness of what we've done, "enough to make your eyes tear", as Meredith puts it. For example, when Europeans first came to America, a squirrel could travel through primary forests extending from the Atlantic Ocean in the Southeast to the Great Lakes and the Mississippi Valley without ever touching the ground. Then the pioneers began to cut down trees that seemed to go on forever, so that they could build homesteads and grow crops. During the nineteenth century, growing cotton and tobacco raised the ante by depleting topsoil and diverting water from its natural flow. During the twentieth century, we have exacerbated the situation by industrializing the landscape with power plants, factories, textile mills, urban sprawl, grid systems and agribusiness in cotton, tobacco, grazing and other environmentally unfriendly practices. Even since the time of William Vogt this past century, the poisoning of the environment has escalated ever more, providing a recipe for disaster if we don't reverse what we're doing. The wilderness experience of centuries gone by has vanished forever. As our memories fade we become increasingly desensitized and our standards drop ever further.

The shift to clean energy is an important symbol for what we must do, but it's only a beginning. In the following sections, we look at what might bring back the most significant layer of the onion: we must preserve and restore our principal ecosystems to sustainability, diversity and beauty. In spite of political gridlock on these issues, it is encouraging to know that there are a dizzying array of solutions, many of which are already being implemented by courageous and innovative individuals and groups in various local areas throughout the world. What follows is a small sample of the kinds of things we can do.

Biodiversity

The massive human-caused extinction of planetary species is escalating in spite of the warnings of biologists such as Harvard's Edward O. Wilson and the Smithsonian's Thomas E. Lovejoy. "Among the repercussions the greenhouse effect is likely to have, the hardest to mitigate is the loss of biological diversity", wrote Lovejoy. "Today biological diversity is signaling that the sheer numbers of people combined with our effects on the environment have almost reached the point of no return."[5] The slight rises in temperature, for example, have driven much of the zooplankton out of the coastal waters of California, the main food source for much sea life, which has gone away. Monarch butterflies that alight on genetically engineered crops often die from the chemicals in the food crop derived from pesticides. Birds of every description flee and die from toxic waste and the poisoning of the atmosphere. Frogs are disappearing or grotesquely mutating.

Meanwhile, we humans seem to be doing almost everything we can to decrease the diversity of our plant species, raising the risk of contaminating food and wiping out natural medicine supply. Humans now consume about 40 percent of the Earth's biological productivity, according to John Tuxill of Worldwatch. "We are eroding the very ecological foundations of plant biodiversity and losing unique gene pools, species, and even entire communities of species forever", wrote Truxil. [5] "It is as if humankind is painting a picture of the next millennium with a shrinking palette—the world will still be colored green, but in increasingly uniform and monocultured tones." One in eight species are immediately threatened.

To make matters more challenging, the genetic diversity of food supply keeps decreasing. For the sake of improving agricultural efficiency, large centralized seed providers narrow the choices for farmers and raise the risk of major crops being wiped out by blights or bad weather. The farmers lose their independence as they are beholden to huge multinational chemical corporations supplying the seeds for every year's crop. The resulting crop contains the chemicals of pesticides with uncertain nutritional implications for billions of consumers. We already know what is happening to the butterflies. Right now in the United States there is a debate about whether genetically altered foods should be labeled as such. Now they are not.

The growing influence of Monsanto and other biotechnology giants provides the means of decreasing crop diversity. The fate of our collective health and food supply could become precarious, according to many accounts in the literature [6-8]. The only recent gain seems to be the creation of crops more resistant to externally applied pesticides. But the risks are great.

"These days we are dealing with...unreported and secret coups-d'etat", wrote investigative reporter Jon Rappoport. "(for example), the promotion of genetically engineered food crops all over the world. The ownership of all the major food-seed companies by chemical corporations is altering these seeds to absorb more pesticide, and give birth to only one crop (terminator technology), forcing farmers to come to the corporation every year to buy new seeds." [6] Rappoport gave numerous examples of the corporate manipulation of media and politicians on this issue.

"In the end," said Tuxil of Worldwatch, "plant diversity can be securely maintained only by protecting the native habitats and ecosystems where plants have evolved." Here is one very compelling reason to leave the remaining undeveloped land alone. How can we not preserve what's left as we restore what we've messed up? This should be an immediate priority number one policy change: *Place a moratorium on further development of any native habitats.* This is a good early project for altruistic private money and enlightened public protection, as in the case of the National Parks. In order to preserve biodiversity, aesthetic value and the quality of life, we will need to buy up such lands for their preservation for all time, or at least for as long as we remain stewards of the planet.

Also needed would be a diverse seed pool for farmers. "Plant diversity (both wild and cultivated) is held mostly by developing countries, but the economic benefits it generates are disproportionately captured by industrial nations....The right of subsistence farmers to save and adapt the seeds they plant still has not been recognized by many governments." Truxil feels that the solution must lie with "part of a larger process of shaping ecologically literate civil societies that are in balance with the natural world." [5]

Food Supply and Agriculture

Not only is the world food supply becoming more uniform and therefore more vulnerable, but the excess use of water, depletion of topsoil, overuse of fertilizers and pesticides, the growing population and the desertification of the planet are all conspiring to make the situation far more perilous if we don't make fundamental changes in our agricultural policies. Land and water are finite resources that have gone beyond sustainability. But intensifying agriculture Monsanto-style is certainly not the way to do it. There are simply too many risks, too many chemicals and too much centralized control. "The main contribution of genetic engineering in agriculture in the future," writes Lester Brown, "is likely to be in the breeding of disease- and insect-resistant varieties. This will contribute to additional production only if these biological pest controls are more effective than the chemical controls now used." [7]

Brown warns that the impressive gains in agriculture during the latter twentieth century are now levelling off. What's more serious is that we have "hit the wall" in the further availability of water because of the threefold increase in irrigation of crops and ranges over the last fifty years. The remaining water is almost gone, as water tables drop and some of the great rivers of the world don't make it all the way to the sea. The soil is also wracked with fertilizers, up nine times since 1950. William Vogt would be horrified with these statistics for sure. Brown argues that we currently don't have a way to feed a growing population.

Fortunately, we can intensify agriculture through more environmentally friendly approaches. One particularly insightful, highly acclaimed and little known book that explores solutions is *Get a Life: How to Make a Good Buck, Dance Around the Dinosaurs and Save the World While You're At It* by Wayne Roberts and Susan Brandum [9]. I met them both on a yoga retreat in Canada and find their information both crucial and timely. In a very major way, this book gives many specifics on the ideas we will need to restore the biosphere and the necessity for a new fellowship of humankind through these trying times.

They cite, for example, an obscure agency called the Office of Appropriate Technology Transfer for Rural Areas (ATTRA). The U.S. Fish and Wildlife Service began this operation when they discovered the great impact on wildlife of the dumping of toxic agricultural chemicals. ATTRA provides advice to farmers to keep their practices sustainable. "Sustainability is a journey, not an end point", says ATTRA's Bart Hall-Bayer. "We're a bridge between the farmer and rapidly changing farming technologies."

On the agricultural subject, Roberts and Brandum explore a broad range of topics, including: community supported organic agriculture (eliminating the need for energy-wasting long range transport); the dangers of agrichemicals; the high ecological cost of eating meat; and the blessings of biodiversity, manure, permaculture and hemp.

Permaculture can literally place the cover back on the onion of lost soil and water (it provides onions as well) by preserving top soil from tilling and from chemical invasions by fertilizers, pesticides, herbicides and genetically engineered crops. Permaculture works best in wooded areas, where a variety of crops can live together in harmony with nature, and where the natural water is not disturbed. "The idea is to orchestrate an ensemble of mutually beneficial or symbiotic relationships", Roberts and Brandum write. (p.125)

The authors profile Jeremy Rifkin, whose book *Beyond Beef* describes the "energy guzzling" cost of meat eating in the world of agribusiness. [10] "To down their 65 pounds of beef a year", said Roberts and Brandum, "North Americans have saddled themselves with a food system at the wrong end of the law of diminishing returns...Cattle trample on a land base that could produce five, 10 or 26 times more of higher quality protein, fiber and nutrients if it were devoted to grains, beans or spinach respectively...A pound of beef takes seven times the amount of water needed to produce a pound of vegetables, rarely noticed thanks to $2 billion in irrigation subsidies to a small number of western ranchers." (p.114)

Sad to say, ever more forests are felled in the tropics to make way for grazing. In the more temperate zones, the land is shrinking by desertification and water diverted from once vibrant rivers and water tables. Meanwhile, huge pig farms in the Midwest and Southeast are springing up. According to the Sierra Club, one of these produces as much sewage each day as the city of Los Angeles.

An extreme example of corporate welfare, local pollution and cruelty to animals is Seaboard Corporation, an agribusiness that grosses almost $2 billion a year. While its pig farms occasionally provide jobs in Midwestern towns whose residents were initially happy to land the business, the company would later move on, leaving behind a mess and an impoverished community, as described in a poignant expose reported in the November 30, 1998 issue of *Time* magazine.

During the 1990s Seaboard netted for itself "economic incentive" payments of over $150 million from state and federal sources. The pigs were inhumanely treated, allowed to die stinky deaths on the farms at the rate of dozens each hour, and collected only occasionally in trucks. The ones lucky (or unlucky) enough to survive are transferred to processing plants—in one location forty thousand hogs are crammed into forty-four metal buildings, where they are fattened for the final kill. The stench is so horrible that people living in the neighborhood have to wear gas masks and close their doors, windows and chimneys. Jobs are low wage and volatile. If the corporate welfare incentive to relocate elsewhere becomes attractive, the company will lay off their people, abandon their mess and move away to another town hundreds of miles away.

Meanwhile the absentee owners live near Boston, raking in hundreds of millions of dollars of private wealth. Is this fair? Certainly not, but sad to say, Seaboard is only one example of how big business is subsidized with our hard-earned tax dollars and then allowed to abuse the sanctity of life. The situation could be ameliorated by ending the subsidy system, taxing the polluters, and providing the incentives where they really belong: for environmentally friendly initiatives.

Now the Good News: Hemp

Imagine a world in which logging takes place only on existing tree farms, and where paper and construction materials are abundant. Imagine replacing petroleum with greenhouse-neutral methanol while we wait for a full-fledged hydrogen or new energy economy to come in. Imagine that the resulting carbon dioxide buffer will help stabilize the Earth's atmosphere against human-caused global warming and climate change. Imagine replacing cotton (comprising fifty per cent of polluting agrichemicals in the United States today) with a crop that needs and leaves no chemicals. Imagine going from dirty petrochemicals to substitute chemicals, oils and plastic made from sustainable biomass. Imagine creating abundant new foods and medicines at low cost.

Quite surprisingly, one substance could fulfill all of the requirements listed above: hemp. It's an old standby that was "as American as apple pie", as reported by author Jack Herer. [11] The story of cannabis hemp is an extraordinary one, going back 2000 years. Even though George Washington and Thomas Jefferson grew it, drafts of the Declaration of Independence were signed on it, the first American flag made out of it, and clothing, sails, oil and rope produced from it for centuries, this versatile crop has disappeared from the American (and world) scene since the 1930s. Why? Because a blend of marijuana xenophobia and the fears by an entrenched timber industry that their monopoly on supplying the world's paper and construction materials might end. Over twenty years, the annual growth from one acre of hemp can provide the same amount of cellulose fiber pulp as woodcutting on four acres of forest during the same period. Herer presents a compelling case of politics and industry run amok, thus blocking a promising possibility.

Most of the xenophobia comes from the relationship between commercial hemp and smoking pot. While they're both from the same family, the consensus is that one would have to smoke a ton of commercial hemp to get even a mild buzz. And as for sneaking in marijuana crops among the hemp, that is not possible: cross-pollination would destroy the mind-altering aspects. So why the big fuss? Industrial self-interest. Fortunately, several states in the U.S. are lifting the long-standing prohibition against hemp. Companies are forming to make use of this miracle crop, which had been unjustifiably blown out of existence before most of us were even born. Using marijuana itself, considered by anti-drug zealots as a top priority illegal drug, results in the imprisonment in America of hundreds of thousands of otherwise innocent people. Yet smoking pot has in fact proven to be more benign than inhaling tobacco and drinking alcohol. Voters in California and other states have approved marijuana as an effective pain-relieving medicine prescribed by doctors.

Once again Europe is ahead of America in its public policies. For example, Dutch scientists have a $40 million budget to research hemp as a substitute for lumber as a paper source. The European Economic Community supports farmers who grow it at a rate of $400 per acre.[9] Several states in the U.S. have begun to approve the legalization of hemp for commercial uses. This amazing crop, in unison with new and renewable energy, may become center stage in the revolution to restore the biosphere. Hemp is a perfect substitute for lumber, paper, cotton, tobacco, petrochemicals and competing biomass. It is gentle on the soil and is agriculturally sustainable after many plantings and harvests. But the sixty-year global hiatus on hemp research and use must be lifted. Then we could produce more robust seeds for a vibrant hemp economy. We can requalify ourselves to re-inherit the Earth only by developing sustainable concepts free of industrial and political bias. Hemp provides a major piece.

At the root of all this is the quest for crops that can be substitutes for products that destroy the environment. Hydrocarbons that come from the ground consist of molecules comprised of carbon and hydrogen. When burned, they create great pollution. On the other hand, carbohydrates that are grown in fields consist of molecules containing carbon and water. When these are burned, the results are much cleaner, especially when we extract hydrogen as fuel. In order to complete the work we need to do, we are going to have to move from a hydrocarbon economy to a carbohydrate, hydrogen and new energy economy.

"Hemp will be the flagship of the carbohydrate economy and the giant killer of (the dominance of the multinationals)", wrote Roberts and Brandum [9]. "It makes food, fuel and fiber accessible almost everywhere on the continent, undermining monopolistic control over scarce resources...It spares forests from unnecessary logging, renders imports and transportation of bulky commodities unnecessary, and converts diffuse, low-level solar energy into high-energy products. But it is not the only boat in the green armada. Straw, for instance, which is now burnt off as waste, will likely come to predominate in paper and manufactured wood, because it can be picked up for prices that will likely undercut hemp as the dominant material." (p.201)

In 1997 I spent a week at the Sivananda Yoga Centre in Quebec staying in the same straw bale dormitory as Roberts and Brandum. The largest of its kind in North America, this building has proven to be less expensive, more insulating, quieter and ecologically more friendly than any wood structure. This is the time when we must dare to think differently.

And then we have Azomite, a remarkable fine powder from Utah that contains trace minerals so often missing in today's depleted soil. Called such by its discoverer Rollin Anderson, it contains the "A to Z of Minerals", and has been shown to greatly enhance the health and yield of fruit trees and crops.   It also enhances the soil for grazing by animals. Quoted in the book *Secrets of the Soil* by Peter Tompkins and Christopher Bird, Anderson concluded this about his discovery: "We now know that Azomite aids the soil caused by mineral depletion or deficiency from continued use over long periods of time. Soil without humus is half alive, and without bacterial action humus is dead. The reason the bacteria in the soil fail to function properly is because of the lack of natural trace elements and catalysts." [12]

In summary, the ideas are there: we just need to have the good sense to support the required research and implementation. There must be a reassessment about how we might be able to safely intensify our agriculture, make our croplands sustainable and improve the quality of food. It is all possible in principle to achieve these things, but we shall have to move beyond the parochial interests of big agribusiness.

Sustaining our Forests

With such substitutes as hemp, we have a chance to make a needed industrial shift from subsidizing the systematic destruction of the treasury of the Earth's forests. Here are some relevant statistics: More than ninety per cent of the forests in the U.S. have been logged at least once, so we have lost most of our great primary forests.

Most deforestation is now happening in the tropics. About sixty per cent of the world's tropical rain forests have been lost already, mostly to slash-and-burn farming, grazing and logging. The rate continues at an alarming half a hectare per second, or one to two per cent of the remaining forests per year—about the size of England. These forests are headed towards extinction well before 2100. When the trees go, so does the topsoil, the cool canopy of the forest, the absorption of carbon dioxide, the retention of water, the prevention of (remaining) forest and brush fires, and preservation of biodiversity, indigenous cultures and natural medicines. The deforestation of Indonesia has been particularly alarming, contributing to out-of-control fires that in 1997 caused such a wicked haze that thousands of people died of lung diseases and a jetliner crashed in the pea soup atmosphere above Sumatra, killing all 232 people aboard.

Interestingly, only certain kinds of forests and crops can efficiently act as a buffer for reducing carbon dioxide in the Earth's atmosphere. A rough estimate made from data compiled by the Institute for

Global Futures Research in the U.K. [13] suggests that planting trees covering an area one-third again larger than our current forest area would be enough to absorb the excess carbon dioxide. When combined with preserved existing forests, these regrowth areas could   restabalize the atmosphere, while we take decisive steps to end the fossil fuel era. So we could develop a strategy to switch to renewables and to plant new trees on clearcut land, overgrazed rangelands, unproductive agricultural land and depleted watersheds. The replanted forests could also provide new habitats and permaculture food sources while mitigating floods, forest fires and extreme weather.   In some limited cases they could become tree farms which can be used for hard-to-substitute wood products and be replanted on a rotating basis. Then we can begin to terraform ourselves back to letting nature handle any climate change. Who, outside the monied special interests, would not be in favor of that?

There's another way to remove excess carbon dioxide from the atmosphere. Oceanographers have discovered that replacing dwindling amounts of phytoplankton algae in the ocean with iron sulfate granules dramatically speeds up the reformation of algae which absorbs carbon dioxide and brings the sealife back to feed. [14] In one experiment the scientists sprinkled 1000 pounds of iron sulfate over a 25 square mile area depleted of phytoplankton. Within two weeks, the new biomass sucked up 2500 tons of carbon dioxide from the air. They estimate that scaling up the experiment to 11 per cent of the world's oceans could zero out global warming over the next thirty years.

While this development could provide an important near term solution to global warming and climate change, it should not give Americans or any other polluter any excuse to ignore the emissions standards set out in Kyoto and the Hague. The Europeans have made it clear that Band-Aid solutions to emissions cannot work in the long run, and must therefore be contained at the source. There are just too many problems with hydrocarbon pollution to rely solely on finding sinks for selected greenhouse gases. Moreover, the feedback loops are too many and too complex to be playing God with Gaia; the only elegant solution in the long run is to cut to the quick of the problem.

Getting back to the task of preserving and replanting our trees, we must become aware that the ownership of our precious remaining forests is widely distributed and subject to exploitation. In fact, only eight per cent of the world's forests are protected. [15] The consequences of continuing current practices are grave, and an international system of protection must exceed the paltry efforts of the United Nations. We will need a world governance to save this rapidly dwindling and essential resource before our planet becomes a wasteland. Former U.S. president Ronald Reagan had said that once you've seen one redwood you've seen them all. This was used as a philosophical justification to chop most of them down, leading to strong support by himself and his successor George Bush Senior for logging interests. Neither man understood the depth of the issue, nor does the present Bush.

Earlier I mentioned that preserving the remaining forests is absolutely essential to restore and maintain biodiversity. Humans must reverse the enormous mass extinction we have sadly created for ourselves. We now have additional reasons for doing that, not the least of which are preventing floods, erosion, desertification, forest fires, droughts, escalating carbon production, and climate change while preserving water, food, lumber, paper, oil and mineral resources. But there is also the quality of life of directly experiencing these habitats in harmony with nature. We shall need to place a value on our forests that would make it prohibitively expensive to cut more down.

We must find ways to stop logging and burning pristine forests and to keep them intact as if our lives depended upon it. (Do you remember Julia Butterfly, who spent one year living in a California redwood to prevent its demise?) We have to make it so that the lumber is not worth getting except on designated tree farms. Mere money or legal or illegal land ownership or use should no longer dictate our future. One time, when we lived in the mountainside woods of Oregon, the absentee owner of the adjacent lot suddenly had the property clearcut while we were away on a trip. The action was devastating to the habitat and to us, this once fairy world of pines, fir, oaks, boulders, moss, roots and flowers. The devastating effects of logging can be seen everywhere, a symbol of our greed and folly.

A number of public and private groups have already begun the process of preserving what's left, yet most of our forests and habitats remain endangered. The positive examples we see now can be seen as a model for what must happen. One group, Conservation International, has flagged down tropical wilderness areas and various "hotspots" where enough forests are still left to protect many species from extinction (*Time*, Dec.14, 1998, pp.64-65). The Nature Conservancy has bought up the Palmyra Atoll in the Pacific and 1.3 million acres of forest in the U.S., while media mogul Ted Turner purchased another 1.7 million acres in the

American West for $500 million (*The Christian Science Monitor,* May 4 and July 25, 2000). But these are just the beginnings of what must be done all over the world.

Most of the remaining pristine forest habitats are in the tropics. Sustainable logging efforts there don't meet with the same success as those in the more temperate areas. Just a few trees like mahogany are desirable. Once those are logged, researchers discovered that the habitat is effectively destroyed. Moreover, the timber harvest operation has opened the land to new roads that are later used for slash-and-burn farming, grazing and local fuelwood (*U.S. News and World Report,* June 29, 1998).

Even more sobering was the Brazilian government's recent decision to pave over the last 435-mile stretch of highway that penetrates the rainforest linking the Amazon River with southern Brazil, putting at risk one third of the remaining dense forest. [16] Succumbing to economic pressures, Brazilian politicians are also considering the release of even more of its remaining rainforest to private destruction and development. So it looks as if there is no easy way out of this dilemma besides a cultural turnaround that would prohibit all stripping of pristine forests. Finding emergency resources to buy up the land becomes more important than ever. We shall need hundreds of times more money than the Nature Conservancy or Ted Turner could muster up. The financial and human resources are there, but we must dare to dream about ways in which private and public moneys could be released. An investment on the order of a trillion dollars now spent on cost-ineffective dirty energy and wars would go a long way to preserve our global forest habitat once and for all. And, by the way, land is not a bad investment.

We must also find ways of renewing land devastated by slash-and-burn food production and logging. We shall need to rethink our agrarian and forestry practices from the bottom up. Let's give the farmer and lumberjack the spade. Let's start planting new trees, as in the case of China described below. Let's keep the woods as source of a quarter of all the world's medicines, a place for limited ecotourism, permaculture, and a heritage of indigenous cultures who might at last thank us for thinking ahead for a change—just in time.

One such culture, the Cogi, has managed to isolate itself from the West by retreating to the far reaches of the high Colombian mountains. We recently met with Alan Ereira, the producer of an award-winning BBC documentary "From the Heart of the World". The Cogi regard the wasteful Western way as the actions of their "younger brothers". The unprecedented drought they've experienced on their land precipitated the warning from the Cogi elders that the Earth is dying and can no longer tolerate the abuse by the younger brother.

A trip I once took up the Amazon from Iquitos, Peru, revealed the effects of deforestation in short order. While the main Amazon was awesome in its size and savannah lands, its banks had been logged. I was particularly struck when we cruised upstream into a tributary surrounded by primary forests. The sheer sizes of the trees and their diversity and the chorus of creatures day and night, the echoing thunder and rain, provided for me an experience of nature whose grandeur I shall never forget. We simply cannot apply power saws and torches to any more of this monumental refuge. Like the world's oil production, which will peak this decade, the world's tropical forest extraction rate is greatest at this somber time. In both cases we have already exhausted about half of the supply, so inevitably, the prices will rise and supply will dwindle—forever.

Even the ecologically crude laws of economics dictate that the prices for our rampage will go up, making even the rape itself a foolish act, when there are so many other ways to do the job of growing food, timber and fuel. Meanwhile, we shall have to find ways to replace local fuelwood for the vast populations of the developing world, so it too can be preserved. The rapid dissemination of solar stoves, new energy and/or hydrogen technology, once available, will greatly support the overall biospheric restoration effort.

We can see what happens in those cases where the forests are cut down to the point where a region or island becomes an overnight wasteland. Easter Island was once a land of rich forest. But its Polynesian culture near the time of greatest wealth ended up chopping down most of its trees. Anthropologists agree that the denuding of Easter Island was probably a primary factor in the decline of the culture thereafter. We also see in modern times the example of the Philippines, when during the 1960s and 1970s, ninety per cent of their forests were cut down mostly for exported lumber. The result was "the nation has become a timber importer and 18 million forest dwellers have become impoverished." [15] Who benefited and what will they do with the remaining ten per cent? The most basic question to be asked of those responsible for unsustainable ecosystems and economics is, why did you do this? Was it for profit only?

Similar shortsightedness also plagues Indonesia, where logging has created massive erosion, forest fires, floods, migrations and killer smogs. A few wealthy individuals made handsome profits in the short run from these practices, but the people paid dearly, and the Earth became stripped and ugly. For better results, we will need to create an enforceable international order to prevent such trashing from ever happening again.

China is beginning to make progress, in spite of its own shortsightedness in preserving its timberlands. Their shift was prompted in 1997 by extreme flooding and mudslides along the Yangtse River which killed thousands of people, affected the health and habitats of hundreds of millions, and destroyed billions of dollars worth of croplands. The disaster could be directly traced to logging 85 per cent of the forests in the upper part of the basin, which then became a desolation of mud that slid down the river.

Once the tree roots had held the fertile soil; now all that is gone. These events grabbed the Chinese government's interest. They halted the logging industry and re-employed one million lumberjacks to lay down their axes and pick up their spades. These new planters have become Johnny Appleseeds to restore the watershed. In one swift action the Chinese were able to begin the job of preservation and restoration. The rest of us could learn a lesson or two from this. Rather than crying out for jobs, we the public need to have some say on what those jobs ought to be in the common interest. The will of the people will have to take precedence over the self interest of companies and wealthy land owners who are in it for exploitation and temporary profit.

To summarize, the overall effort to save our forest habitats and resources has to move in many directions, with the same sense of urgency that we approach the questions of energy, agriculture and water use. The new directions include:

1. *Preserve what's left of all natural habitats and forests by buying up the land and by placing a moratorium on logging, slash-and-burn farming and fuelwood recovery, aside from tree farms;*

2. *seek ways of replacing wood with hemp, straw, bamboo, free and renewable energy;*

3. *restore damaged habitats by planting trees and working more wisely*

with water; and

4.   *continue protecting , restoring and sustaining habitats so they may be preserved for all time.*

An impossible job? Yes and no. Our focus now is to consume ever more resources and to make our money grow faster and faster. We must therefore outlaw our destruction of Gaia at all significant levels and reallocate those resources to restoration. We must preserve what's left and invest in what will clearly become our most precious resource: the awesome bounty and diversity of Earth.

Our mantra ought to be, preserve, restore and sustain.

As a species, we will need to create a moral shift in attitude so we can organize and get on with the job. This will be the focus of the next chapter.   Our immediate priority should be to buy up the endangered land. Then place a high value on it and keep it for current and future generations. But we must also provide support for anyone who might be displaced in the process.   We can't be distracted from these goals and actions, or ultimately we are doomed. Necessity is the mother of invention.   This time the invention is not deploying some new physical technology of energy; it is a simple act of will and reverence in shifting our priorities. To summarize:

> *Preservation* involves a financial investment by those who could afford it to buy up and preserve our remaining forest and other identified "hot spot" natural habitats before it's too late. We must allow nature to run its course in those regions.

> *Restoration* will involve investments of trillions of dollars, but create jobs in the areas they are needed most. Following the example of China is one way to do the job. Logging and farming interests suddenly become conservation and replanting interests.

> *Sustainability* should once again be easy: let nature keep doing its thing and let's keep making Earth a better place to be stewards of. Let's preserve these precious places for our survival

and quality of life, so the Earth itself doesn't become another sorry example of an Easter Island, Philippines or maybe even Mars or Venus.

The Magic of Water

I write this while I am visiting Dennis Weaver's Earthship in Colorado on a hot, parched day in early August 2000. The grass is uncharacteristically brown, the streams are drying up, only tiny blotches of snow appear on the north slopes of fourteen thousand foot peaks, and the seasonal dazzling colors of the alpine wildflowers show up only in wilting patches near high streams. Some afternoons we are greeted by a thundershower that might produce a drop or two of passing ecstacy. Forest fires rage in nearby Mesa Verde National Park, blotting out the usual blue sky.

Yes, global climate change is upon us as the jet stream streaks across northern Canada far away. Even during a normal, moist summer, the water here is manipulated and coaxed over to cattle rangelands, mines, golf courses, crops and growing cities downstream.

According to the recent research of Lester Brown and others, a major obstacle to sustainability and adequate food is our dwindling water supply. We have already diverted to our food-growing areas almost everything we can from our once-mighty rivers, whose mouths have become silty, muddy and sometimes dry. Water tables are dropping rapidly in many significant agricultural areas. Indeed we have "hit the wall" on all this: global fresh water consumption has tripled since 1950. The truth is, both the quantity and quality of our water have degraded at alarming rates in recent years. We have on our hands another mega-problem demanding a mega-solution, before it's too late. The trends are ominous. Peter Phillips, Associate Professor of Sociology at Sonoma State University, cites plans by Monsanto, Enron and Bechtel to buy up water supplies throughout the world (www.projectcensored.org). As in the cases of oil, natural gas, food and lumber, these corporate giants antici-pate windfall profits in the face of increasing scarcity. Governments are signing away their control over domestic water to these companies. This privatized takeover of another essential resource we had taken for grant-ed adds a new dimension to the problem of giving away natural capital and to bribing the "regulators". *The control of our natural resources must be vested in the public.*

The water challenge has many added subtleties that run deeper, as the ever shallow rivers, lakes and reservoirs continue to lose their purity and vitality. The layers of the onion are many on this one. First we have the eroded and desertified land as the supply becomes unsustainable. We have hydroelectric dams that can create ugly man-made lakes, displace millions, keep rivers from flowing in a natural and healthy way, and provide little in the way of control to counteract floods due to human-caused climate change. We have a polluted and unsustainable system of aquifers worldwide that needs cleaning and redirecting in flows that more accurately reflect its former natural state. I remember in my youth swimming in the clean rivers and lakes of New England, and dipping my Sierra Club cup into the streams of the high mountains in the West for a refreshing drink of clean water. You can't do either now without inviting stomach diseases from grazing animals upstream. Most of our potable water is treated and sapped of its nutrients in order to be drinkable. Bottled spring water can cost more than the same amount of petroleum or beer, an unheard-of prospect while I was growing up.

This mess we've created moves right out into the oceans, where both the supply and diversity of sealife are dying off, principally from overfishing, toxic dumping and global warming that kills coral and phytoplankton. Ocean liners illegally dump garbage, supertankers keep spilling oil, and the Navy blasts sonar off Hawaii, driving away and possibly wiping out whales, dolphins or anybody else that would be in harm's way. Thanks to a quintupling in harvesting fish since midcentury, once again we are on the sharp summit of production, already beginning to take a nosedive permanently beyond sustainability. This high extraction of seafood will soon dwindle as are also the world's remaining petroleum, trees and available water. Coral reefs are dying and the once biodiverse waters off California have become a desert because of the sensitivity of microorganisms in the food chain to slightly warming temperatures. Again, we have "hit the wall" in many major respects, a signature of our frighteningly unprecedented exploitation of the Earth.

Meeting these challenges of both quantity and quality of water habitats and water use requires our most urgent attention—now! The human desertification of the Earth, the declining quality and availability of water, and the preservation of global aquaculture are each in themselves major issues to be urgently investigated as public policy, to be addressed by a global green democracy described in the next chapter.

The Quality of Water

Even though we're familiar with its chemistry and many of its unique properties, water is magic stuff . All life as we know it depends upon liquid water to survive. We are made mostly of water. We need it to drink. Water covers two-thirds of the Earth. It has subtle, life-giving qualities. By the way that we've been redirecting, tapping and polluting our water, we seem to have little sense on how vital this resource is to us and to the web of life. Because of its abundance, we have taken water for granted.

Recent hikes in the Colorado Rockies have made it clear to me that water running down a natural stream is bubbling, singing, happy. Water running through human-dug canals seems brackish or restlessly racy. Pipes and dammed-up lakes are ugly and also disturb the natural flow.

Early in the twentieth century, the Austrian forest warden Viktor Schauberger (1885- 1958) made some important discoveries about the behavior of water in nature. He noted that cool water running in natural streams with a forest canopy overhead provides conditions for the water to have many life-giving qualities that can deteriorate and become diseased if the water then runs through man made flumes, canals, pipes and hydroelectric dams.

By observation and controlled experiments Schauberger found that water needs to spin and swirl the way it does in natural streams to be healthy. These findings transcend ordinary chemical changes, although even those can happen too, because water can transform itself to greater purity when it is spun. My own experiments with the late Marcel Vogel verify that fruit juices passing through a spiral tube surrounding a specially charged quartz crystal can hold floating mold for months without any reaction with the mold, whereas the control samples become immediately contaminated and blackened. [17]

The basic properties of water that give it this magic quality are not well understood by mainstream scientists but are nonetheless real. In Chapter 6 we will look at some extraordinary experiments that water has a memory and can respond to our intention. Stanford University materials scientist William Tiller has successfully conducted experiments in which the acidity and various other qualities of water can be altered through human intention transmitted by an electronic device containing the memory of that intention. [18] In the previous chapter, we have seen in cold fusion experiments that hydrogen atoms within water or heavy water at room temperature can be enticed into reactions that produce new nuclear materials and the release of large quantities of energy, in seeming contradiction to traditional nuclear physics. Alternatively, the Mills hydrogen gas cell appears to produce anomalous energy when heated up in the presence of a catalyst. The clean fuel and exhaust in both cases is water.

Several definitive and replicated experiments show that various homeopathic medicines can become biologically more effective at very high dilutions of water, such that not a single molecule of the medicine itself remains in the solution. [19] Somehow water responds to, and maintains, the subtle energies of consciousness coming from nature and humans. There is something within water and its constituent hydrogen that might provide the medium for a science and ecology of the future.

Author Callum Coats brings up to date many of the ideas of Schauberger, who late in his life fifty years ago (like William Vogt) sternly warned about the implications of our escalating neglect of the environment. [20] Schauberger's simple prescription: *"imitate nature."* This important principle contradicts the prevailing paradigm of seeing water as simply a solution in which to dump waste and sewage, to treat chemically and to divert in gross ways. So few of us even in the mainstream environmental movement appreciate the seemingly more subtle nuances of water which nature produces and humans destroy.

For example, Schauberger showed that straight or angular canals and flumes draw the water in a nonvortical (nonspinning) motion that measurably depletes the quality and purity of the water. To verify this, I have conducted controlled experiments with Vogel in which dowsing and tasting yielded clearly better results with the spun batches.[17] Hydroelectric projects further ruin river water when it slaps against the turbine blades and loses its invigorating oxygen content. Metal and concrete pipes don't provide the natural materials necessary to carry pure water. Canals and pipes for irrigation do the same thing, thus contaminating ever further our food.

Coates describes Schauberger's concepts of having an adequate water drinking supply. "Apart from our own consumption of it," writes Coates, [20] "this same water is also used to grow everything we eat. If we want to live in health and happiness, then the living entity—*water*—should be highly revered and the most sensitive care taken of it." (p.193)

Unfortunately, present methods of water treatment kill the water and often include adding poisonous chlorine and fluorine, thus decreasing our resistance to disease. Schauberger has found ways to repurify the dead water through designing various configurations that imitate nature. But to get to the roots of the problem, we will really need to expand this science which has been so neglected, as we continue to insist that water is only water—to be grossly manipulated and treated chemically. Who cares, we have been led to believe, what form water takes, as long is it's filtered or chemically purified or used as a dumping ground or wrung out of the land, rivers, lakes and oceans?

The symbiotic relationship between water and our precious remaining forests cannot be overemphasized. "Without photosynthesis", said Coates [20], "we could not survive, so our continuing existence is wholly dependent on this great gift of oxygen that only trees and other vegetation can provide. Were there no trees there would be no animal life, human life or (the wide variety of) micro-organic life on this planet. When trees are cut down indiscriminately, we not only harm them, but we harm ourselves as well, for by doing so we reduce the amount of oxygen and water available to us." (p.210)

Regarding the continuing decimation of forests and increasing use of prematurely harvested, uniform regrowth forests, Coates wrote this: "Rotation is reduced to an absolute minimum and biologically speaking represents a denial of the future, because no tree is allowed to reach full maturity. It is a process akin to killing a child. While the age of a mature redwood is about 2,000 years, today it is felled after 60 years of growth. This means that it has been cut down when only 3% of its full potential has been realized and before it can be fruitful. As an act of violence it is equivalent to slaying a human being with a life expectancy of 70 years when it is just over 2 years old. As a result there is no longer any mature seed and gradually the genetic base of the seed deteriorates to the point of infertility. The consequences of this madness are far-reaching, for as the biological diversity is depleted of its highest quality organisms, so too are the qualities, energetic and otherwise, that support higher forms of life. The destruction of the forest goes hand in hand with the destruction of water, and...the consequences of this insanity are appalling." (p.226).

Schauberger summarized the problem: "If the forest dies, then the springs will dry up, the meadows will become barren and many countries will inevitably be seized by unrest of such a kind that it will bode ill for every one of us." [21]

Regarding our insensitive approaches to agriculture, Schauberger said: "Contemporary agriculture treats Mother-Earth like a whore and rapes her. All year round it scrapes away her skin and poisons it with artificial fertilizer, for which a science is to be thanked that has lost all connection with Nature." (quoted by Coates, p. 261). It is sad to ponder that, since both Schauberger and Vogt made these kinds of statements in midcentury, worldwide fertilizer use has gone up ninefold.

Clearly we are going to need to become more educated about so many aspects of our destructive ways swept under the rug in the name of profit and (inadequate) pretenses toward sustainability. It may not be enough to be able to recycle our aquifers. For water is the lifeblood of all nature, and has many mysteries to reveal if only we take a look before its essence in nature vanishes. The grossness with which we are treating these rapidly dwindling supplies even goes beyond hunger, climate change, and insisting on providing a growing stream of products from trees.

We must now question the existing complex infrastructure of water-engineered projects to produce energy and food. And we must question our reliance on traditional chemistry and filtration to do the job of purification. The web of life depends upon water's abundance and its quality, and the solutions we must seek may seem daunting, but not so much so if we've already begun to look at what we've been considering so far in the areas of energy, agriculture and forestry. This particular concern feeds back on all the others, and demands sophisticated solutions matched with the best of our intellect and respect for nature, our heritage and our future.

I am also aware of some avant garde research on water, described on the Internet and elsewhere, which promises to purify our waterways once and for all. [22] I haven't yet seen the demonstrations, but would not be surprised to see positive results coming in as the research expands. The old infrastructures will now need to be questioned and new ideas considered in the way we handle and use water before we can even begin to requalify ourselves to reinherit the Earth. We need vigorous research programs to test the theories of Schauberger and others, and to find ways of restoring and sustaining the world's waterways. We need to conserve, recycle and repurify our water, and to stop contaminating it with fertilizers and pesticides. And we must look at the potential of desalinizing sea water as a water source in a new energy economy. As in the case of the fossil fuel energy debacle, we will need to consider innovative solutions to the water dilemma we have created, just now coming to light. It is interesting to note that the magical properties of water itself might provide an important answer to the challenges of having abundant energy, food and forest resources, while keeping air, water and land pure and natural.

Recycling and Waste Disposal

Last but not least in our plan to restore the biosphere is our conspicuous consumption and waste of raw materials. I see it every day, the plastic and paper bags and packages thrown away into expanding land fills, the energy intensity of packaging, transporting and disposing of new garbage, and the more troublesome problem of toxic waste which pollutes ever more our precious waterways. Two Yale University economists estimate that the average Westerner consumes irrevocably about $100,000 of natural resources, about 30 per cent of the Gross Domestic Product of those nations. [23] This does not include the effects of global warming, climate change, or ozone depletion.

It is perhaps symbolic that one very lethal technology that could still kill us all—nuclear energy—has been used for the peaceful generation of electricity, but with the very heavy price of deadly radioactive waste materials which can stay in the environment for tens of thousands of years before transmuting themselves into benign substances. This challenge has met with no end of problems, as Federal projects try to deal with this (hidden) cost of nuclear power. In the United States, a solution has been found, although not an elegant one: high level radioactive waste is now being dumped into deep caves in Nevada. The cost has escalated into tens and billions of dollars, with no end in sight for further environmental and economic costs of further dumping. The spent radioactive fuel comes from materials in nuclear power plants that have given us a small percentage of electricity, in addition to the production and stockpiling of doomsday weapons which make no sense. With the radioactive burial, what would happen with a major earthquake or water event that would alter the crust of the Earth? Once dumped, it's out of our hands. What about future generations who have to live with the stuff long after we die? Isn't our overindulgence a bit crazy? How might Gaia herself feel about this injection of long-lived poison?

There seem to be more elegant ways to deal with the problem, including the possibility of transmuting the wastes using cold fusion

technologies mentioned in the last chapter. Early experiments are showing positive results in remediating radioactive and other toxic byproducts
into benign substances. This would be an elegant solution. At another
level, we have the enormous social cost of toxic waste in our waterways,
oceans, mines and land fills. These measures have created an enormous
number of public health hazards which are underpinning the most vigorous arm of the environmental movement: lawsuits by aggrieved individuals against the polluters. In the end, the solution must lie in creating
a global jurisdiction for stopping the pollution and recycling waste—not
only in lengthy and expensive legal proceedings at local levels.

Disposing of nuclear and toxic waste is the sensationistic tip of a
nightmare of an iceberg. Sooner or later, we shall have to recycle everything. In America most estimates point to the fact that we've taken some
measures to bring recycling up to 10 percent, but the goal will need to be
100 percent. I am not proud to be a citizen of the leading waste producer. The United States usurps a third of the world's consumed resources.
On the average, each of us throws away more than our own weight every
day: 101 kilograms! [24]

But the energy companies and individuals are not the only
wasters. "Mining has contaminated thousands of kilometers of rivers and
streams in the United States alone", write Gary Gardner and Payal
Sampat of Worldwatch, "and logging threatens vital habitat, often of
endangered species. Air and water pollution from manufacturing plants
have sickened millions, often shortening lives. Some of the 100,000 syn-
thetic chemicals introduced this century are a ticking time bomb, affecting the reproductive systems of animals and humans even a generation
after initial exposure. And the effort to make waste disappear—by bury-
ing it, burning it, or dumping it in the ocean—has generated greenhouse
gases, dioxin, toxic leakage, and other threats to environmental and
human health." [24]

The attitude of extracting, possessing and throwing away so
many materials which we feel is our birthright is fundamentally incorrect. The new perception will require radical changes in our consuming
habits to go to a recycling economy. In the end, we must insist that all
materials used be sustainable. We the offending consumers are going to
look at this one straight in the eye, ranging from recycling everything to
not "owning" such a variety of things like lawn-mowers that are only
occasionally used by one family rather than shared. We have been so
steeped in valuing and accumulating things. But the world situation

demands that that pattern be reversed.

"Recognizing the absurdity of our materials-intensive past", write Gardner and Sampat, "is a first step in making the leap to a rational, sustainable materials economy. Once this is grasped, the opportunities to dematerialize our economies are well within reach. Societies that learn to shed their attachment to things and to focus instead on delivering what people need might be remembered 100 years from now as creators of the most durable civilization in history."

Reflections and Summary

I reflect on the depression I feel about the changes we will have to make soon. It will almost certainly demand a quick shift of attitude towards what has value. It will necessitate a radical lifestyle change in the United States, particularly about waste management and recycling.

Yes, I recycle bottles and paper. But the garbage still goes out on Fridays bound for the land-grabbing, polluting, smelly landfill which feeds our water tables with ever more poison and produces ever more of the greenhouse gas methane. I feel guilty about my own responsibility in accumulating and throwing away so much stuff, of adding my own part in putting more carbon dioxide into the atmosphere every time I turn on a switch, fill the bathtub or turn over my car engine. Even printing this book, though on recycled paper, was an agonizing moral decision, yet well-researched in terms of its small overall environmental impact as opposed to taking fossil-fueled trips to advocate the same things.

The point is, we must first restore sustainability to the Earth. The corporations aren't doing it, the governments aren't doing it, as we keep slipping away from the goal and hit the wall in the supply and quality of our greatest treasures. So the rest of us will need to do it. The technological and ecological solutions I propose here are not difficult to understand:

1.   *Shift energy sources to emissionless new and renewable sources.*
2.   *Preserve, restore and sustain our forests, grasslands, crops, waterways and oceans.*
3.   *Develop new agricultural options such as hemp, straw, azomite, organic farming and permaculture as substitutes.*

4.   *Re-examine the role and use of water for health.*
5.   *Convert the materials economy to a 100% recycled one.*

Meeting these challenges may seem to be socially insurmountable, but they must be met; the logic is so clear. Oddly enough, it is the fifth area, recycling our resources, that might meet with the greatest resistance from most people in terms of changing our lifestyles—especially in America.

I realize that much of this may appear as a threat to the goals of the existing infrastructure and shatter the temporary illusion of a prosperous standard of living. Population growth exacerbates the pressures. New social entities may need to act in authoritarian ways at times, and the transition won't always feel easy. We shall look at these questions in the next chapters. Meanwhile I hope I have convinced you that very real physical things are happening on this planet that demand physical solutions—before it's too late!

On the positive side, the ecological mandate will not only deliver us a planet back in balance. It will also trigger revolutions in our thinking. As we clean the planet we shall also be able to clean our minds and hearts to re-discover the great beings we really are. Happenings on the frontiers of science, medicine, and spiritual matters don't give us only ecological hope. They will literally transform our attitudes and empower us towards a new civilization. The concluding chapters will discuss this good news—our personal and planetary paradigm shifts and expanding knowledge about what it means to be human. These activities blend very well with our efforts to restore the biosphere.

References for Chapter 3

1. Lester R. Brown and Christopher Flavin, Chapter 1 in *State of the World 1999*, Norton, New York, 1998.

2. Palden Jenkins, Healing the Hurts of Nations, Numbers 91 and 92, www.isleof avalon.co.uk/palden, 1999.

3. Albert Gore, Jr., *Earth in the Balance: Ecology and the Human Spirit*, Houghton Mifflin, New York, 1992.

4. Lester R. Brown and Jennifer Mitchell, Chapter 10 in *State of the World 1998*, Norton, New York, 1998.

5. John Tuxill, Chapter 6 in *State of the World 1999*, Norton, New York, 1999.

6. Jon Rappoport, *Notes on Scandals, Conspiracies and Coverups*, Truth Seeker, San Diego, 1999, p. 48.

7. Lester R. Brown, Chapter 7 in *State of the World 1999*, Norton, New York, 1999.

8. Vandana Shiva, *Stolen Harvest*, South End Press, Cambridge, MA, 2001.

9. Wayne Roberts and Susan Brandum, *Get a Life: How to Make a Good Buck, Dance Around the Dinosaurs and Save the World While You're At It*, Get A Life Publishing House, Toronto, Canada.

10. Jeremy Rifkin, *Beyond Beef*, published in 1992 and quoted by Roberts and Brandum. [8]

11. Jack Herer, *The Emperor Wears No Clothes*, AH HA Publishing, Van Nuys, CA, several editions from 1985 to 1998.

12. Peter Tompkins and Christopher Bird, *Secrets of the Soil*, 1998.

13. Institute of Global Futures Research, Global Futures Bulletin, www.rofa-cfan.org, 1999.

14. Philip W. Boyd et al, "A Mesoscale Phytoplankton Bloom in the Polar Southern Ocean Stimulated by Iron Fertilization", *Nature*, vol. 407, pp. 695-702, 2000.

15. Janet N. Abramovitz and Ashley T. Mattoon, Chapter 4 in *State of the World 1999*, Norton, New York.

16. Eugene Linden, "The Road to Disaster", *Time*, October 16, 2000.

17. Brian O'Leary, *The Second Coming of Science*, North Atlantic, Berkeley, CA, 1992.

18. William A. Tiller, *Science and Human Transformation*, Pavior, Walnut Creek, CA, 1997.

19. Michael Schiff, *The Memory of Water: Homeopathy and the Battle of Ideas in the New Science*, Thorsons, 1998.

20. Callum Coats, *Living Energies*, Gateway Books, Bath, U.K., 1996.

21. Viktor Schauberger, "The Dying Forest", *Tau Magazine*, vol. 151, Nov. 1936, p.30, and quoted by Coats.

22. Examples include Johann Grander's Water Revitalization from Austria and Ayhn Douhuk's Perfect Water from Turkey.

23. William D. Nordhaus and Joseph Boyer, presented to the February 15, 1998 meeting of the American Association for the Advancement of Science and reported on *CNN*, Feb. 23, 1998.

24. Gary Gardner and Payal Sampat, Chapter 3 in *State of the World 1999*, Norton, New York, 1999.

PART II

Social and Personal Action Steps

"In (the place of philosophy) rises an empire of stone, steel, smoke and hate —a world in which millions of creatures potentially human scurry to and fro in the separate effort to exist and at the same time maintain the vast institution which they have erected and which, like some mighty juggernaut, is rumbling inevitably toward an unknown end. In this physical empire, which man erects in the vain belief that he can outshine the kingdom of the celestials, everything is turned to stone. Fascinated by the glitter of gain, man gazes at the medusa-like face of greed, and stands petrified."

-Manly P. Hall

NO PARADIGM CAN be shifted purely by having new technologies or declaring the biosphere unfit for habitation. We are going to need to agree on how we implement our plans responsibly, to make the transition as painlessly and compassionately as possible. To purely rely on big corporations or big government to set our policies and priorities has proven to be disastrous to the environment and taxing to personal freedom. We are going to have to create stronger but more flexible institutions which will navigate us through the turbulent waters of change.

Our new direction must begin at home. Communities sharing resources and talents could form nodal points for communicating globally, over the net and through travel, new ideas and green technologies that can benefit everyone. But we are going to need to re-define what is meant by value in this social creation we call money and the economy. We must now factor in the irreplaceable value of our ecology and not only sustain it. We need to give back to the Earth by restoring it, to whatever degree possible, to its former pristine state.

And we must restore government to the people, operating more on local levels as well as globally. We need founding mothers and fathers to forge a Declaration of Interdependence...to democratically elect a rotating Council of Elders who would be empowered to implement the new paradigm with deep love and understanding of all humans and all nature. We must also eliminate from the Earth our weapons of mass destruction. We need to look at all viable options to reduce our impact and population, so we can sustain our home, our one and only beautiful

Earth.

We will need to form a global green republic with jurisdiction over our shift to sustainablility. We must probe more deeply into how our underlying social malaise could have allowed us to destroy our precious habitat. For us to organize ourselves in such a way that the few temporarily prosper and everybody else suffers, needs to be examined carefully. The malaise is cleverly diguised in the media and can be overcome only by transcending economic tyranny.

In this part, we also look at the new science of consciousness as a guiding principle of enhancing our inner and outer ecology. We learn about our great progress in integrated East-West medicine and mind-body models of health and healing. We examine the development of our spiritual selves through the practice of yoga, meditation and other experiential approaches. We investigate research on our immortal selves that seem to survive the deaths of our human bodies. We cite case studies that inspire us to re-examine the possibility that we have a soul which goes on through all time. And we look at the personal meaning of our discoveries and interactions with non-human intelligence, whether they be with aliens, angels, sea mammals and other animals, nature spirits or Gaia herself.

These new models of our transpersonal experience of the seemingly ineffable can open remarkable gateways to our greater being. It is encouraging that these activities are not only available to our own experience and demonstrations: many are also open to experiments using the scientific method. With these personal developmental tools in hand, we can begin to explore territories where the physical meets the metaphysical, where the visible meets the invisible, where science meets the spirit and where our consciousness and the interconnectedness of all things become the ground of all being. Then our greater purpose for being human on Earth at this time can be revealed. Then we may re-qualify ourselves to re-inherit the Earth.

4

Form a Global Green Republic

"To sin by silence when they should protest makes cowards out of men."
-Abraham Lincoln

NOW WE COME to the hard part: how can we humans come together in global community to fulfill our ecological mandate? This step has been most difficult to take because we seem to have so much trouble giving up vested economic and political power for the higher good. It is one thing to become aware of a physical reality and its sustainable solutions (undoing the deplorable condition of the environment), and quite another to take responsibility for unseating those now in control and restoring democracy.

You may not agree with some of my views of this chapter which explore the dark side of American politics and economic globalization. Many of my well-intentioned colleagues have suggested I avoid politics entirely, because it would offend some of my readership and limit marketing opportunities for my expression. They feel I should only look for solutions, for unity, and somehow the human obstacles will take care of themselves. But in this time of great peril, we cannot divorce political and economic reality from what must be done, try as we might otherwise. Our problem is more a human one than a technological one.

American author Charles Reich blames our inaction on what he calls economic government.[1] Large faceless corporations manipulate public policy to their own benefit while placing obstacles in front of any change. But change we must, for our survival. Reich recommends that we must oppose "the System" and reassert our civic responsibility. "In the long run", Reich concludes, "we are going to need a science of social change. We have applied scientific knowledge to virtually all of the practical areas of life except government and economics—two areas still dominated by myths and ideology. If our economic and political institutions become dysfunctional, we will need to create a new area of knowledge to deal more intelligently with human affairs." (p.6).

It amazes me, for example, that the fossil fuel industry, politicians, and media all speak with one voice in allowing for greed to prevail during the 2000-01 California energy crisis, in which customer prices skyrocketed, suppliers hauled in windfall profits, the grid system became overstressed, people lost their power, and the pollution continued unabated. The question of renewable and new energy, from the President and Secretary of Energy on down, was suppressed from public discussion—even though these sources could provide the ultimate answer to the California and global energy crises. We only heard about the crises themselves and not the solutions. One of President George W. Bush's top energy advisors and campaign funders was Kenneth Lay, CEO of the now infamous energy giant Enron*, which has made handsome profits on California's natural gas consumption. Lay was just given a bonus of $ 7 million for his contributions to the corporate bottom line.

The unraveling of a sustainable future seems assured as long the administration led by George W. Bush, who in early 2001 took office as 43rd President, is still in power. Bush is deeply ensconced within the System. He and Vice President Dick Cheney are oilmen with some of the worst public environmental records in American history. They are wasting no time to support the interests of big business to the detriment of a sensible ecological future for humankind. In their first few months in office, they have rebuked the Kyoto agreement, cut funding for renewable energy, and reversed their only environmental campaign promise: to curb emissions. They have convened a secret energy task force whose ambitions are to build new fossil fuel and nuclear power plants, to ease restrictions on pollution, and to drill for more scarce oil in the U.S. West and Alaska. The greed and collusion with big corporations in today's America are greater than they have ever been. These actions could spell big trouble for our future, but it could also provide opportunities for starting an ecological revolution.

"The government becomes infected with the profit motive", says Reich; "the corporate sector uses governmental power without constitu-

*please see end of chapter for footnote

tional restraint. Both resort to lies and deception and whenever possible avoid candor with the public. Both seek to shift blame to someone else... corporations are no longer responsible to their owners, and government is no longer responsible to the sovereign people." (p. 67)   Even the American Government itself looks increasingly like a huge corporation.

Reich states : "We must reject the tyranny of economics so that it restores the human habitat, obeys the laws of nature, and offers a secure place to all members of society...It is time for a restoration of vision. What we need beyond anything else is a new map of reality." (p.24)

Reich recommends that we oppose economic tyranny, rampant pollution, voter disenfranchisement, a free market that is not really free but benefits a management elite, an artificial division between "public" and "private" sectors which are really one and the same at top levels of power. He examines the myth that economic growth is essential, that individuals are responsible for rising within the System, that their success depends upon it, that they need no protection from it or from poverty, that welfare is wasteful, that overcoming social pathology is an individual responsibility and not a societal matter, that the two political parties represent a full spectrum of views, that the courts are fair and the final arbiter of disputes, and that we need no essential change.

Presidents throughout American history have warned us about the political crisis which is now upon us. Two centuries ago, President Thomas Jefferson said: "I hope we shall crush in its birth the aristocracy of the moneyed corporations, which dare already challenge our government to a trial of strength and bid defiance to the laws of our country."

Sixty years ago President Franklin D. Roosevelt said we must establish "practical controls over blind economic forces and blindly selfish men." Without such controls, technology would be a "ruthless master of mankind." Then he told Congress, "The first truth is that liberty is not safe if the people tolerate the growth of private power to the point where it becomes stronger than that of their democratic state itself. That, in essence, is Fascism..." [2]

In 1961 outgoing Republican President Dwight D. Eisenhower also warned Americans about the excessive and undemocratic power of what he called "the military-industrial complex." Or what about another Republican President Abraham Lincoln's quote in the 1860s at the beginning of this chapter? What would Mr. Bush say about these statesmanlike declarations? Would he agree? Almost certainly not, because the U.S. government is now pursuing a path towards protecting rather than balancing the interests of big business.

The prophecies of several past presidents appear to be fulfilled under the Bush administration. His Cabinet choices are hostile towards the environment. Secretary of Interior Gale Norton was a protege of the infamous James Watt, President Reagan's Secretary of the Interior, who opposed the Endangered Species Act and supported logging, mining and oil drilling on public lands. Along with Bush, she has spoken out for drilling in the Arctic National Wildlife Refuge. As a former lobbyist for the lead paint industry and as former Colorado attorney general, she defended land owners' rights to pollute. She joins Mr. Bush in rolling back former President Clinton's public land protection initiatives.

Former Senator Spencer Abraham, Bush's choice for Secretary of Energy, has opposed the development of renewable energy and fuel efficiency programs as "not having returned a very good investment for the taxpayers." Research and development budgets are being cut for these programs. He was one of only three senators to introduce legislation to abolish the very Department of Energy he now heads. He too supports oil drilling in the Arctic and continued coal strip mining practices that contaminate water. In 2001, Senator Frank Murkowski of Alaska introduced a bill to subsidize the oil industry $21 billion to drill there—an outrageous act of corporate welfare! The only bone the bill threw to the environment was to allow clean cars with single passengers to travel in carpool lanes. How many of these absurd bills will pass is unknown at this writing. But the overall effect of America's 2000 coup d'etat seems to be moving ahead more rapidly than many of us could have ever imagined, making the precepts of this book stand in poignant contrast.

The new head of the U.S. Environmental Protection Agency, Christie Whitman, has no significant environmental credentials besides a traditional Republican attitude of public noninterference with private polluters. Still, I can admire her sincere attempts to fulfill Mr. Bush's only environmental campaign promise to limit emissions from power plants, and to find ways to fulfill the Kyoto agreements. But Bush's later reversal of his pledge ended up causing her (and the U.S.) great embarrassment with respect to international colleagues.

Secretary of Agriculture Ann M. Veneman who once worked for Bush's dad in his Department of Agriculture, now oversees the expanding exploitation of public lands including logging, oil drilling and road building in national forests. She has also publicly advocated the widespread distribution of genetically modified foods.

The contemporary question is, can we trust these people to meet in secret and "re-engineer" an electrical generating future which will affect all of us on this planet? My answer to this action is a firm "no" and begs the question that we need a global governance to limit the excesses of private industries and their sympathetic governmental collaborators.

With the appointment of practicing corporate ideologues rather than defenders of the public interest, this indeed is a dark day in American affairs. At this writing, while Mr. Bush presses ahead with his dangerous agenda, the oil tanker Jessica spills hundreds of thousands of gallons of highly toxic diesel fuel into waters near the pristine Galapagos Islands off Ecuador, and the most recent estimates of global warming by IPCC scientists have soared to as much as ten degrees by the end of this century.

There is even the question of George W. Bush's legitimacy in office. Although his opponent Al Gore won the national popular vote and probably the deciding Florida vote, the Bush legal team successfully petitioned the U.S. Supreme Court, which in a highly partisan 5-to-4 decision, awarded the Presidency to Bush by stopping a crucial vote recount in counties with poor voting equipment and large black and Jewish populations. Just after Bush "won" the election, I saw him on television getting out of his gas guzzling sports utility vehicle on his Texas ranch with a smirk on his face. I sensed that the Bush team epitomized the power of the System. His first agenda items are tax cuts that would benefit the most wealthy Americans and opening public lands to private exploitation and pollution. The question we need to ask is, how much damage can these two men do to our democracy and to our land while they're in office? Or could a miracle or disaster change their minds?

In light of the bizarre and incredible events of late 2000, noted senior historian, author and ambassador John K. Galbraith, by no means a Green, has gone so far as to say that our political system now mimics the very large corporations themselves which now dominate public affairs. In an article "Against Amnesia: Corporate Democracy; Civic Disrespect" widely circulated on the Internet, Galbraith says, "The United States left behind constitutional republicanism, and turned to a different form of government. It is not, however, a new form. It is, rather, a transplant, highly familiar from a different arena of advanced capitalism. This is corporate democracy. It is a system whereby a Board of Directors—read Supreme Court—selects the Chief Executive Officer. The CEO in turn appoints new members of the Board. The shareholders, owners in title only, are invited to cast their votes in periodic referenda. But their franchise is only symbolic, for management holds a majority of the proxies. On no important issue do the CEO and the Board ever permit themselves to lose."

The reaction to the American election fiasco from the press has been mostly bland, but there is the occasional jewel. Particularly interesting are some reactions from abroad, where the thinking tends to be more independent. For example, we have this reaction in England from Will Hutton in the December 24, 2000 issue of The Observer, entitled "Right-wing Coup that Shames America": "What has happened was beyond outrage", writes Hutton. "It is the cynical misuse of power by a conservative elite nakedly to serve its interests—and all of us should be frightened of the consequences ...The incontrovertible abuse is that Bush has won power despite losing, and critically he only pulled off this feat because the Republicans control the Supreme Court."

Overcoming our Economic Tyranny

In my quest to understand the extent to which economic tyranny has overwhelmed an immobilized public into ignoring the solutions of Part I, I highly recommend the works of David C. Korten, formerly on the faculty of the Harvard University Graduate School of Business. His 1995 book *When Corporations Rule the World* [3], Korten lucidly describes the alliance among a powerful minority of economic rationalists, market libertarians, corporate management and their financial backers who have successfully advanced an ideology of capitalistic greed as the "highest expression of what it means to be human." This school promotes the myth that "the relentless pursuit to acquire and profit leads to socially optimal outcomes and that it is in the best interest of society to encourage, honor and reward these values." [3]

Korten's credentials as a scholar in economics and business well qualify him to address these issues. He sees the large corporation as "an alien entity with one goal: to reproduce money to nourish and replicate itself. Individuals are dispensable." One is reminded of Star Trek's alien culture The Borg, which systematically assimilates individuals for its own cancerous fulfillment. The increasing power of the cartels, says Korten, "has seriously debased the integrity and social utility of economics...that violates its own theoretical foundations...It is creating a world that they would scarcely wish to bequeath to their children." (Ref. 3, p.73)

Korten documents that this "monster" we have created has placed the rights and freedoms of corporation ahead as those of individuals: "Presented as an economic agenda, it is, in truth, a governance agenda." For example, the 2001 relaxation by the U.S. Federal Communications Commission regulation of media conglomerates and recent federal court decisions are now granting the rights of these corporations to consolidate their power, as if they were to have the same free speech rights as individuals. The result is the ever decreasing diversity of public opinion on the airwaves, in the newspapers and in magazines. Korten continues: "It's almost as though we were being invaded by alien beings intent on colonizing our planet, reducing us to serfs, and then excluding as many of us as possible." (Ref. 3, p. 74)

Korten's case studies, statistics and conclusions are chilling reminders that could lead reasonable or thoughtful people to exercise their rights as public citizens to end the tyranny before it's too late. Korten states that we as a culture have defaced the positive essence of the proper role of Adam Smith's merchantilism, consisting of small buyers and sellers in balanced trade. We have instead created the very same kind of monopolies and cartels which Smith would have abhorred.

There is an accelerating tendency for large corporations to Merge, Acquire and Dominate (MAD) on a global scale. These actions are clearly leading our species to dis-inherit the Earth. In this chapter we look at the need to create an aware and enlightened world democracy to stem the tide of unbridled globalized economic power. We must move beyond the assumption that we are competitive, separate, selfish units in a world ruled by ruthless giants. The human species can only re-inherit the Earth through unprecedented social organization which will ensure we are part of a greater sustainable system, unified in diversity.

"The more dominant money has become in our lives," writes Korten, "the less place there has been for any sense of the spiritual bond that is the foundation of community and balanced relationship with nature...the task of people-centered development in its fullest sense must be the creation of life-centered societies in which the economy is but one of the instruments of good living—not the purpose of human existence. Because our leaders are entrapped in the myths and reward systems of the institutions they head, the leadership in this creative process of institutional and value regeneration must come from civil society." (Ref. 3, pp. 6-7)

Korten's second book *The Post Corporate World* proposes solutions outside the box of corporate capitalism and globalism. His model is to look at sustainable living systems based on cooperation rather than at mechanistic systems based on competition. To build future models, I believe we are going to need to look beyond synergistic living systems and into the spiritual meaning of our interconnectedness and the results of holistic science as significant ingredients for solutions. A new science of consciousness may provide the needed foundation for acknowledging greater truths for which neither mechanistic economics nor biological systems provide adequate models—although living systems are surely more wise than our corporate rulers.

On a practical level, Korten's arguments are compelling. He urges that capitalism cannot be globalized without a counterveiling democracy. Of great importance here is the principle that the power of no economic system can exceed checks and balances by public participation, which recognizes the tyranny of economic globalization and the unity of all life on Earth. Korten provides many near-term solutions to our dilemma: regulate corporations and money markets on an international scale; phase out the World Trade Organization, the International Monetary Fund, and the World Bank; restore political democracy to nations; end corporate campaign financing; end corporate welfare; break up large corporations and help small ones; help communities, families and workers; prohibit, limit or tax speculation, leveraged acquisitions and mergers; adopt local currencies; and restore legally the rights of personhood over propertyhood. These measures are good candidates for the global Simultaneous Policy mentioned in the Introduction.

In my book *The Second Coming of Science*, I showed that we have to put ourselves outside the box of materialistic Western science before we can begin to glean the significance of the new scientific revolution we are in the midst of—despite the violent objections of the mainstream scientists supported by the government, universities and media. The same goes for the economic government itself: one must step outside it to begin to understand its tyranny. I join Reich and Korten in believing that in taking this small step, we can expose the System for what it is, free of the captivating bromides such as "free trade", "economic growth", "prosperity", "end big government", "no new taxes", "compassionate conservatism", "liberal spenders", and "personal responsibility" which can be very deceiving or outright lies. Even the word "conservative" is a misnomer because most conservative politicians conserve nothing natural. Rather, he or she preserves a dysfunctional System antithetical to any kind of conservation measures.

Among Reich's proposals is the concept of "Gross Domestic Cost" for polluting. The new discipline that will have to supplant ordinary economics is what Dennis Weaver calls ecolonomics—an economics that cherishes our natural heritage. We need a pollution tax, and the place to start would be to shift public subsidies and research funds from the polluters to green initiatives and new energy. Corporate welfare of the polluting kind is one of the most egregious assaults on our collected common sense and must end. Two places where you can find a restless public eager for social solutions are recent demonstrations and the international Green movement.

The 1999 World Trade Organization (WTO) protests in Seattle and the emergence of Ralph Nader as the U.S. Green Party candidate for President are recent examples of opposing the System from outside its confines. The WTO is a "private" organization acting as an unaccountable government to look after the interests of big business and the wealthier nations in providing them trade opportunities worldwide. This arm of the System drew Gandhi-like demonstrations by thousands of concerned citizens outraged by the rapidly growing power of the economic elite embodied in the WTO. For the most part, the U.S. media tried to paint the protesters as "mindless anarchists" and "violent leftists", without asking what might have motivated the protests in the first place, or the fact that they were, for the most part, intelligently managed and peaceful. An example of media cynicism is Charles Krauthammer's essay "Return of the Luddites" published in the Dec. 13, 1999 issue of *Time* magazine. The title says it all.

The myth of "free trade" embodied by the WTO can often be an excuse to pollute in Third World countries which, under the pressure of debt, must deliver natural resources to their financial masters abroad. "Deforestation", writes Alex Falconer, Member of European Parliament, "is the inevitable result of the rapid economic growth policies adopted by many developing nations in response to the demands of the global financial institutions." [4]

"The (unsustainable) growth imperative permeates all the problems we face. It makes us look at the world through a distorting lens. Productivity is measured in terms which ignore pollution by toxic and climate changing chemicals. When modern agriculture, based on fossil fuels and chemical fertilizers, uses ten calories of non-renewable energy to produce one calorie of food, is it really an advance upon traditional practices that yield two calories of food for every calorie of renewable energy input?"

The transnational corporations, says Falconer, are more powerful than whole nations and "can even take on and win in battles against supranational institutions such as the European Union. The supposed regulators of international trade and finance, such as the WTO, are in practice captives to the Transnationals' agenda." [4]

Falconer is looking for activists to support the need to globalize ecology as well as trade. The local pool from which he could draw in Scotland includes U.N. Non-Governmental Organizations (NGOs), churches and labor groups. Along with the environmentalists, these are the kinds of people who coalesced for the WTO demonstrations in Seattle. Such coalitions, in addition to other new thinkers, will form the nucleus of a robust global green democracy which would oversee the conversion to an ecolonomic and just future.

American Politics and the Green Party

During the 2000 U.S. presidential campaign, Green Party candidate Ralph Nader challenged the American Bush-and-Gore mainstream positions, which ignored the most important issues of the environment, labor conditions, universal health care, military spending, growing inequalities between the wealthy and others, campaign financing reform, crime prevention, accelerating imprisonment of citizens, minority rights, and judicial and legal system abuses. Both major party candidates seemed like robotic, well-paid actors running on non-issues and vying to grab a bland political center. So as a result, they managed to split themselves right down the middle in votes cast by a bewildered and sometimes disenfranchised electorate. The poetic ending of Bush's victory in the Supreme Court overtime was most interesting but frightening. Many of us still have withdrawal symptoms from watching this CNN television cast 37-day sporting marathon.

I make no apology: I am a Green and I voted for Nader (Appendix II). Like any political movements, it has its faults, but the Greens' philosophy is basically sound on public issues. We must remember that Nader got three per cent of the national and Florida votes— enough to make a difference in the 2000 election and enough to grow on. All this happened with barely any funding, media coverage and his exclusion from the presidential debates. Nader quadrupled his votes from 1996 to 2000; what might happen in the future? It is also heartening that it also took only about three per cent of Americans during Revolutionary times to muster the courage to oppose the tyranny of the king. It doesn't take a majority to incite change, but it will to consummate it in a functioning democracy. For all its faults, the U.S. will soon be joining the Europeans in activating the green movement. This will become necessary for us to form a global green republic to oversee the changes we will need to make. I use a lower case for "green" so as to include not only Greens but other groups which will need to make up the new worldwide coalition.

I estimate that we have a few years to plan a new system. During that time we can watch the Bush administration do its thing for everybody to see, spurring us on and stirring us up in the process. We can learn more about the essence of the System directly from one who practices it to the hilt. We can then view what must be transformed into something of social value. The Democrats (lesser of the evils in my view) can lick their wounds, rethink their priorities, and plan on being in power. It would be wise for them to talk with Nader and the Greens next time, or they might lose again.

If you've had the courage to digest the essence of Part I, you cannot escape the fact that the System is so deeply flawed it is destroying our environment. We must therefore expose it with the ridicule and irreverence it deserves, and then organize outside of it to build a new system, some of it from the ground up. Then we can ask them to give up their power, forgive them and go on. The System is like a Ship of Fools whose faceless skipper could be thought of as a drunken sailor who somehow got at the helm of Spaceship Earth. The skipper is now steering the ship into the kind of harm's way that befell the Titanic, the Exxon Valdez, the Jessica, the Kursk, the Namouri and the U.S.S. Cole.

In the process of our winner-take-all capitalism, we may have defeated communism and connected the world to a Disneyland of franchises offering unrenewable consumer delights while the band plays on. We can thank the skipper for his contributions and gently replace him with a new skipper whose priorities will be very different. During the coming years we will need to create a new system. In the following sections I sketch what that system might look like and what its first projects could be.

Charles Reich has made some suggestions about what a new system might incorporate. First, we need to ask, What is an economic system for? Whom does it serve? "We have the power to design a system that will do whatever job we ask of it", he writes. "We are not limited to models of the past." (ref. 1, p.194)

I agree with Reich that the System functions like a machine which focuses on the means of production and consumption, far removed from the value of what it is like to be a human being living on the Earth. He cites Lord Tennyson, who wrote: "Our little systems have their day: They have their day and cease to be." [5] We must now have the courage to step outside the box and see the System for what it is, and then move on to the solutions.

Reich's second remedy addresses the question of regaining control. We don't want to work within the System ever again nor do we want to incite a violent revolution. We need a middle ground in which we can create a new system that is "posteconomic", protects citizens from economic tyranny and reactivates our role as citizen-shareholders, following the ideals of democratic individualism in the tradition of Emerson, Thoreau, Whitman and John Dewey.

Third, he says, it is time to renew the U.S. Constitution. Not only has it been abused by the System; many things have changed over the past two hundred years that could not have been envisioned by the framers. The tyranny has moved from the monarchy to economic government, so we will need a new bill of rights to apply to reverse private control of resources, property and the workplace, and the right to clean air, pure water, climate stability and decent housing. To amplify on Reich's ideas, we shall also need a Declaration of Interdependence, acknowledging that we must come together in the global village to declare the need for a green governance structure over matters of common concern to the entire planet. No one national government could do that job.

Reich concludes that we shall need a new social contract with the people. Managers still in power must become responsible to provide economic opportunity to anyone who seeks it and to restore meaningful work and generous allowances for time spent on individual human development. Unfortunately, the interests of citizens and the interests of the global community for a clean environment have not been well served under any recent political system, whether it be capitalism, socialism or communism. We will need to build a new system that moves outside of any rigid adherence to any one of the three political philosophies. Rather, we can choose to take the best of each and create structures that are entirely new.

Finding the Money to Begin

As we begin to come together and move into these systemic changes both in the U.S. and worldwide, we have the practical matter of finding the resources now to re-inherit the Earth. It might take a while to form a global green republic, but money needs to be pried loose now to pay for the most urgent projects. We have seen that at least some hundreds of billions of dollars will be needed to invest in our remaining pristine land, another tens of billions to research, develop and implement a clean and renewable energy economy, and unknown trillions to restore the biosphere: loggers, farmers, contractors and oil workers displaced in the process could be re-employed to grow trees and hemp, for example.

So where will the money come from? First, we have private sources of trillions of dollars. It should come as no surprise that most of the money is locked up in the System itself, because billionaires are System-bred and have used it prudently to rise to the top. They will be among the last to change. Yet we need to find wealthy people who may have had a change of heart, who could see, for example, that the purchase of vast quantities of land would be a great investment for both themselves and the Earth.

In the public sectors, we first see what could happen if we the people and our representatives in government decide to shift subsidies. In America, the "green scissors" reports coming from Friends of the Earth and other environmental groups proposed cutting over $50 billion in 72 federal programs that support research which benefits polluting industries. The green scissors project has more recently been expanded to include proposed cuts coming from tax breaks amounting to hundreds of billions of dollars a year for environmentally unfriendly big business— often more than these companies pay in taxes. These privileges stand in stark contrast to the burdens of personal income taxes, borne mostly by the middle class. [7] Finally we have more hundreds of billions going into provocative and wasteful military projects such as "Star Wars" space weapons favored by the Bush administration. These funds could be converted to green projects, especially now that the Cold War is over.

But there could be even more potent sources of funding from outside the United States. It is likely Mr. Bush will veto any Congressional initiatives to shift subsidies and taxes, so we may need to look abroad in the near future. One source of funding could come from the so-called

Tobin tax, a concept which is particularly popular in Europe. In 1972 the American Nobel Laureate economist James Tobin proposed that a tax be placed on international financial transactions which would set some limits on unproductive and unstable currency speculation (the so-called "Global Casino"). "Some $1.3 trillion changes hands each day in foreign currency transactions. The suggested Tobin tax of 0.25 per cent would raise about $250 billion a year—more than five times the current level of all international debt." [6] Not only could these funds be used to balance foreign debts; they could also pay for a renewal of the biosphere on a global level. If approved in Europe, for example, the tax would transcend national boundaries and be a convenient means to fund the new projects.

Towards a Global Green Republic

In the long run, we will need a global government with certain specific powers to oversee the necessary changes. Exactly what those powers will be is open to debate, but certainly one of its first missions would be to restore the Earth to sustainability. Arne Naess, the founder of the "deep ecology" philosophy, pointed out that the international green movement is comprised of three movements: the peace movement, the social justice movement and the ecological movement, including the goal of ecological sustainability. He also concluded that sustainability should be our first priority because it is most urgent for our survival. [8]

A future global green government would be entrusted to convert our resources to sustainability, prosecute international criminals now protected by rogue nation-states, reallocate massive military and polluting industrial budgets towards a planetary cleanup, protect human rights, and probe a deeper understanding of ourselves and our place in the universe, free of the cultural and intellectual biases of the times. The book *PlanetHood* by Benjamin B. Ferencz and Ken Keyes, Jr. sums up the situation: "We all have the right to live with dignity in a healthy environment free from the threat of war."[9] The fast way towards destruction, they say, is nuclear catastrophe and the "slow" way is our current polluting path—which is not so slow right now.

What would a progressive international green platform look like? I suggest five areas of focus which also summarize some of the views we have looked at up to this point:

1.   *Ecological sustainability and beauty:*
        As we have seen, this broad-based goal will require an all-out
        Apollo program to preserve what's left, to restore what's been
        destroyed, and to sustain and beautify the biosphere for as long
        as we are stewards of the Earth. Just what is meant by sustain-
        ability and beauty will have to be debated among world citi-
        zens and their representatives in global government.

2.   *A lasting and enforceable world peace and world government:*
        Following the prescription of *PlanetHood*, we could replace the
        anarchy of warring nation-states and of rampant industrial pol-
        lution with a new   representative world government which
        would assure us of a peaceful and friendly coexistence and a
        clean environment. A U.S.- and European-led Green move-
        ment well underway could provide the needed impetus. We
        may also want to draft a Declaration of Interdependence which
        would spell out the laws of ecology and what the relation of
        humans to the natural world should be in a sustainable global
        civic society. Such a manifesto (Chapter 7) could open up more
        spiritual approaches such as those of U.S. Natural Law Party
        presidential candidate John Hagelin and sociologist Paul Ray's
        public surveys that nearly one-fourth of the American popula-
        tion consider themselves as "cultural creatives" breaking out of
        our current two-party system [10], which has shamefully resulted
        in 55% of eligible voters choosing not to vote. The cultural cre-
        atives would include a green aspect and a "can-do" attitude in
        visioning a positive future.

3.   *Social justice and fairness to all:*
        History teaches us about the importance of equal rights for all.
        Prejudices regarding race, gender, creed, economic position, and
        lifestyle preference have always played a disproportionate role
        in political and corporate decision-making. The WTO protests
        in Seattle vividly point out the dark side of free trade and glob-
        alization of business opportunity. With corporate mergers to
        help, the rich are getting richer and the poor are getting poorer.
        The obvious results are declining labor conditions and environ-
        mental destruction, particularly in the Third World. The inter-
        national green movement will need to embrace these overlap-

ping issues and provide the means to control excessive economic greed. We will need strict antitrust measures.

4.   *Ecolonomic conversion from pollution, war and economic government
      to green initiatives.*
      During 1967and 1968 I helped Democratic presidential candi-
      date George McGovern with defense and aerospace "economic
      conversion" platforms that would provide for re-training and
      research and development on green initiatives, symbolically
      turning swords to plowshares and factories to Edens. These
      concepts developed more fully in my work with candidates
      Morris Udall in 1975 and Jesse Jackson in 1988. While conver-
      sion policies never took hold politically, they could be dusted
      off, revised, and provide a significant role in creating new
      opportunities and public-private initiatives. Under a green
      government, we could immediately shift subsidies and tax
      breaks from polluting industries and massive military projects
      to more benign efforts such as clean energy development and
      wilderness protection. These policies could go a long way now
      to reorder our ecolonomic priorities, while preserving economic
      opportunities for all: new jobs! As in the case of China, lumber-
      jacks can become tree planters.

5.   *Research and development on solutions such as free and renewable
      energy, hemp, healing, consciousness and the purification of water.*
      Current R&D priorities in the U.S. Department of Energy, for
      example, focus on fossil fuels and nuclear power. But through
      public decree we will need to shift federal subsidies and R&D
      support from polluting power to clean power. Traditional
      renewables such as solar, wind and the promising hydrogen
      economy have been given some support, but not nearly enough
      to become competitive. New energy research on cold fusion,
      hydrogen gas cells and zero point concepts might provide an
      ultimate clean solution to the energy crisis, but have been
      denied and suppressed at every turn. As a public project, it is
      time to research every viable option free of the biases of existing
      economic and political structures. We need to clean house.
      Even the green movement could fall into the trap of opposing
      "far out" possibilities; therefore we still need to have a broad-

based, continuing and open debate of energy-environmental policies. [11]

Let's say that none of the new ecolonomic projects could go ahead in the U.S. and other nations because of System gridlock. Then we can choose to levy some sort of Tobin tax and fund the projects from offshore. There is no reason why we need to deal with the System, including the U.S. government itself—except for the enforcement of pollution control. We have the talent and innovative skills to bring in people and funds to do the job. It will probably be better to employ qualified world citizens who have been outside the System for some time, so that the new efforts won't be tainted or diluted. "There is enough for everybody", says Peter LaVaute, founding president of Ecosense Solutions in Columbia, Missouri. "It is not a matter of scarcity, but one of resource management and distribution." A fellow board member of Dennis Weaver's Institute of Ecolonomics, LaVaute suggests many of the same projects and policies outlined here. "We need to develop a positive action plan to bring the human community in harmony with the Earth...an organization must be created." [12]

Dennis Weaver foresees an Ecolonomic Revolution. "The Industrial Revolution was energized and propelled forward by competition...more often than not, the cut-throat variety. The Ecolonomic Revolution does not discard competition, but realizes the more powerful and efficient energy of cooperation. With the Industrial Revolution, waste is inevitable, a natural part of doing business. With the Ecolonomic Revolution, waste is eliminated by design. The Industrial Revolution is linear in its thinking...the Ecolonomic Revolution is cyclical in its thinking...what goes around comes around. It mimics nature's eco-system and creates Ecolonomic technologies...The Industrial Revolution holds the philosophy that the business of business is business. The Ecolonomic Revolution understands that the role of business is more than just business (read busyness). Realizing it is the strongest, most influential institution on the planet today, the Ecolonomic business understands its role in the larger whole. It is concerned for the community, the workers, and the general welfare of society." [13]

It seems that the long term social solution for a troubled world is to blend global green republic with ecolonomic business. Then everyone would win. But how can we do this? How can we transcend the awesome power of big business and create a new global entity responsive to civic society?

The People are There to do the Job

We tend to forget that we are much larger in number than those who command the massive resources of giant corporations and politics. Paul Ray's cultural creatives include about fifty million Americans and seventy million Europeans who could form a broad-based coalition whose potential influence could eclipse the darkness of our times. We are simply not sufficiently organized yet.

"In the United States", writes American business analyst Paul Hawken on the Internet, "more than 30,000 citizens' groups, non-governmental organizations, and foundations are addressing the issue of social and ecological sustainability in the most complete sense of the word. Worldwide, their number exceeds 100,000...This new sustainability movement did not start this way. Its supporters do not agree on everything—nor should they—but remarkably, they share a basic set of fundamental understandings about the earth, how it functions, and the necessity of fairness and equity for all people in partaking of its life-giving systems...it is spreading throughout this country and the world. No one started this world-view, no one is in charge of it, no orthodoxy is restraining it. I believe it is the fastest-growing and most powerful movement in the world today, unrecognizable to the American media because it is not centralized, based on power, or led by charismatic white males."

So the people and money are there; we only need to do it. But most of all we need to *believe* we can do it. The Internet and jet travel can accelerate the process of coming together in global community to organize and implement sustainability. But as we shall see in the next two chapters, two further elements could provide the needed impetus: the first is to pay attention to our own personal ecologies (in body, mind and spirit). The second is to expand our human knowledge base now being suppressed by powerful vested interests. We must move beyond materialism and support the birth of a new science of consciousness which reflects a unity of all creation (Chapter 6).

What We Must Do

Sometimes it seems I could only wish that a global green republic could naturally emerge to take its proper authority without resistance and without fear of the consequences of opposing the existing paradigm. Not only do the vested interests not want this to happen, but the rest of us will need to muster the courage to move ahead. The new leadership will need to avoid the fallacies of their human egos, which too often seem to block the collective vision. Divide and rule carries the day, and we are deeply afraid of failure. Better the devil we know than the devil we don't know, we think. But we have already seen we cannot afford to have that happen, and so the sophistication of the new social organization will have to be far greater than that of any project in recorded history. This includes the highly organized corporate oligarchy itself, whom we will have to oppose. Our mission is unprecedented. As never before, we must cry out for global justice united in common purpose. We need a global green democracy.

Making the changes will involve a number of innovative steps. They will almost certainly include the following: the launching of world forums on sustainability; the development of the sciences of social change and ecology to help chart our course; the creation of alliances and coalitions among citizens' groups, nongovernmental organizations, foundations, churches and cultural creatives worldwide; and the temporary governance by elders. In our deliberations we cannot be bound by the assumptions of the old paradigm. I imagine we will need to begin our work independently of virtually every old institution with agendas— including even the U.N. Organizations such as these can be used to help with data and experience to assist in the overall effort, but we shall need to begin the political and civic deliberations from an entirely fresh perspective. So here are the first steps:

1.   *Create world forums on sustainability.*
        Here we discuss and debate how we are to implement the goal
        of global sustainability from its ecological roots to its practical
        implementation. Make no mistake: there will be jurisdictional
        conflicts with the interests of large corporations and some
        national governments. But the growing 120 million of us will
        need to point out the consequences of abusive human actions
        and the necessary loss of power among the ruling confederacy
        of corporate oligarchs. We have to trust that sustainable values
        will prevail in the end.

        One example is the World Social Forum held each year in Brazil.

"The World Social Forum", writes the American author Noam Chomsky on the Internet, "offers opportunities of unparalleled importance to bring together popular forces from many and varied constituencies from the richer and poor countries alike, to develop constructive alternatives that will defend the overwhelming majority of the world's population from the attack on fundamental human rights, and to move on to break down illegitimate power concentrations and extend the domains of justice and freedom. The World Social Forum will be a new international arena for the creation and exchange of social and economic projects that promote human rights, social justice and sustainable development."

Interestingly, this group meets at the same time as the World Economic Forum in Switzerland in late January. As a graphic counterpoint the latter group represents what Chomsky calls "the emerging system of 'corporate mercantilism', with decisions over social, economic, and political life increasingly in the hands of unaccountable private concentrations of power, which are 'the tools and tyrants of government,' in (former American President) James Madison's memorable phrase, warning of the threats to democracy he perceived two centuries ago." The contrast between the current exercise of power and the future development of social justice can be clearly viewed when looking at these concomitant forums.

Another organization concerned with sustainability is the State of the World Forum cofounded by Mikhail Gorbachev and American attorney Jim Garrison. I spoke at the fall 1999 San Francisco Forum, which turned out to be quite a star-studded extravaganza of new thinking. The Forum is now focused on invigorating its Commission on Globalization. "The Commission", writes Garrison, "will...explore the critical role global governance must play if the human family and community of nations is to co-create an equitable, just, non-violent and ecologically sustainable future." (www.worldforum.org/commission)

Prince Charles is another world figure who has advocated a sustainable future. In the concluding speech at the annual Reith Lecture Series, he said, "We will never achieve sustainability without a rediscovery of the sacred." Charles proposes a mix of scientific solutions (such as ending greenhouse emissions) and intuitive approaches which respect nature rather than overwhelm her—for example, going to organic farming instead of genetically modified foods. These words stand in stark contrast to the silence of royalty steeped in the control of the world's money and resources.

While I happen to be on the side of radical change versus gradual reform, we will need to create more gatherings like those described above to discuss remedies. No doubt the debates in the forums on sustainability will spawn many factions within the overall movement. This should not detract us from our overall vision so long as we continue to agree on our mission of sustainability. For example, one faction might prefer a moderate approach such as providing a level playing field for sustainable technologies by subsidizing them, while ignoring the politics of pollution. Another faction might say we don't have time for all this, that we shall need to pass and enforce strict new standards set up to protect us from all pollution, and that we should begin immediately with an Apollo program to develop new energy and to preserve, restore and sustain the biosphere. For this we will need to be savvy in the ways of debate, be open to new ideas, and be all-inclusive in civic participation. These will be the town meetings upon the commons of the global village. We will need to adjudicate between differing views and check our egos at the front door of these meetings, never wavering from the basic goal. I mentioned in the Introduction one excellent idea proposed by John Bunzl of England called the Simultaneous Policy (www.simpol.org), where groups within the electorate vote only for those leaders who adopt measures which would be implemented when the issue comes up for consideration.

We must develop a consensus on what is meant by sustainability and to create an action plan of future measures, including budgets and timelines for getting it implemented. (Most of the activity will be at a local level and vary according to regional needs, but have basic global guidelines such as acceptable levels of emissions). We shall need to define the structure of the new democracy by reforming existing constitutions and creating a new draft constitution for the world.

The global commoners could add additional goals to that of sustainability, such as ending hunger, regional pollution, adverse labor conditions, war, weapons proliferation, human and animal rights violations, secrecy, and suppression of new knowledge. Sustainability pulls these other values as well, and they too could become part of the emerging democracy's new agenda.

2.   *Expand the Sciences of Social Change and Ecological Sustainability.*
Here we adopt Charles Reich's suggestion as well as defining what is meant by sustainability, so that we can find a set of

guidelines and monitor the progress of the sustainability movement. We must chart a successful course, in much the same
way as engineering projects such as the Apollo program were
carried out. Ecologists Paul and Anne Ehrlich address the overall lack of scientific knowledge about ecology among the general populace in their recent book *Betrayal of Science and Reason.* [14]
"Most Americans", they write, "readily grasp the issues surrounding something familiar and tangible like the local dump
site, but they have considerably more difficulty with issues
involving genetic variation or the dynamics of the atmosphere.
Thus it is relatively easy to rally support against a potential
landfill and infinitely more difficult to impose a carbon tax that
might offset global warming." So the power vacuum becomes
filled by greed.

Yet the ecologists themselves are eager to take a more active role
in deciding policy. An example are the IPCC climatologists who have
predicted global warming and climate change. *We would now ask scientists
to take these additional steps: (a) formulate sensible (nonpolitical) guidelines
which would limit emissions sufficiently to reverse human impact and restore
sustainability, and (b) analyze the wide range of solutions, including all clean
and renewable energy options.* From this they could construct future scenarios from which the global green republic will have to choose. Social
scientists who have studied paradigm shifts will also be needed; we
could use the wisdom and insights gained by those who have looked in
depth at these things. Arne Naess, Marilyn Ferguson, Barbara Hubbard,
John Naisbitt and Paul Ray come to mind.

3.   *Form Alliances and Coalitions Among Existing Citizens' Groups,
      Local Action Groups, Church Groups, Foundations, Non-
      Governmental Organizations and Cultural Creatives.*
      This step is essential, and must be implemented in such a way
      as to be all inclusive of their own agendas and assure their par-
      ticipation in the new governing body—such as adopting the
      Simultaneous Policy. Some local groups might want to imme-
      diately secede from any global organization because of the his-
      torical record of the tyranny of centralized power. Like technol-
      ogy and economics themselves, the forces from above can act as
      a mindless machine whose understanding of issues impacting

local environmental needs are insensitive and counterproductive. We cannot let that happen: we could throw out the baby with the bathwater making such generalizations about government. The new democracy would have to be carefully checked for any excesses and corruption. It would be designed that way.

4. *Shift Power to Local Governments where Possible.*
The global green democracy and Simultaneous Policy would be limited in their charters to provide resources and guidelines to implement a sustainable future, and related issues of public health, rights and justice. But the locals would make most of the final decisions as to how that would be done in their regions. Only where there are infractions such as landowners' perceived right to pollute or where local governments misappropriate their resources would an enforcement team step in. The basic idea here is to take the power away from "the global greed hypocrisy" and put it into the hands of the global green democracy and local businesses and governments. In this post-industrial scenario, the influences of nation states and large corporations would disappear.

5. *Elect a Rotating Council of Elders.*
Because we are all so wary of the corruptions of power, we will need to be vigilant about who we choose to run this new democracy. I would suggest limiting terms of office to no more then a few years—including the judiciary, and no career posts. I also suggest forming shadow governments and polling constituencies as the new jurisdiction forms. And, please, no campaign donations or voter fraud! Let's take the best from the American and other democracies and leave the rest behind. As to who should lead, I have learned that most wisdom and innovation comes from elders. In my work on new energy, for example, I have identified over a dozen individuals over seventy years old who can lend extraordinary expertise to the effort. Many of these people are professors emeritus from universities whose administrations and faculty are often critical of their work and would just as soon see them go away because their ideas oppose current academic orthodoxy. But because they

don't have long careers ahead of them, they are free to express new ideas with a refreshing level of intellectual honesty, vast experience and unbounded enthusiasm. These are probably the only people I can trust as my mentors, and they could become the mentors of change.

Towards an Ecological Ethic

"The meek shall inherit the Earth."

-Jesus of Nazareth

All major political movements have been based on philosophical underpinnings called world views or paradigms. The ecological ethic is no exception. It is also important to understand and define what is meant by sustainability from the most basic possible scientific and philosophical perspective. This is called deep ecology. Not since the capitalist, social-ist and communist movements of a century or more ago have we been confronted with such potential change. We have already explored in the first chapter that the so-called Information Age touted as the next wave by most of our futurists is but another materialistic enterprise that might improve the efficiency of communications but does nothing substantial to transform the human and terrestrial condition. "I believe that electronic information technology", wrote psychologist Ralph Metzner, "is only the latest, most abstract, expression of the mechanistic, technological mindset and does not represent a real shift in values, such as ecology and the envi-ronmental crisis demand."[15]

"The philosophy of deep ecology," Metzner continues, "teaches biocentric or ecocentric values, in which humans are seen as part of nature, not over or against it....the most radical deep ecologists are not even comfortable with the stewardship notion, since it still implies that humans have superior ecological knowledge and are therefore entitled to take care of the earth." Very humbling. Regarding our current material-istic philosophy, Metzner says, "Global citizens of a unified world in cat-astrophic transition cannot afford to hang on to the fragmentary para-digms of European industrial culture....Technology, instead of being used to feed a runaway cycle of exploitation and consumerism ("more and more for more and more"), will need to be redirected toward the protec-tion and restoration of damaged ecosystems."

Theologian-ecologist Thomas Berry has pointed out that while we have moral teachings for homicide and suicide, they are lacking for biocide and geocide. Mary Evelyn Tucker and John A Grim, editors of the anthology *Worldviews and Ecology*, have put the situation this way: "Our ethics have remained largely anthropocentric (the belief that only humans count) and indifferent to the fate of the natural world. This is changing gradually as we re-examine the nature of human-Earth relations and begin to build the basis for sustainable life in the future...a new global environmental ethics will be needed...we will not preserve what we do not respect." [16]

Tu Wei-Ming, Harvard University professor of Chinese history and philosophy, has traced our current rape of the environment back to the Enlightenment mentality of the eighteenth century, "fueled by the Faustian drive to explore, to know, to conquer, and to subdue (which) persisted as the reigning ideology of the Modern West...this unprecedented destructive engine has for the first time in history made the viability of the human species problematical...We need an ethic significantly different from the social Darwinian model of self-interest and competitiveness. We must go beyond the mentality that the promise of growth is limitless and the supply of energy is inexhaustible. The destructiveness of 'secular humanism' lies not in its secularity but in its anthropocentrism...The crisis of modernity is not secularization per se but the inability to experience matter as the embodiment of spirit...A key to the success of this spiritual joint venture is to recognize the conspicuous absence of the idea of community, let alone global community, in the Enlightenment project....The dichotomizing of matter/spirit, body/mind, sacred/profane, man/nature, or creator/ creature must be transcended to allow supreme values such as the sanctity of the earth, the continuity of being, the beneficiary interaction between the human community and nature, and the mutuality between humankind and heaven to receive the saliency they deserve in philosophy and theology."[17]

Even Al Gore had something to say about our exploitation of the biosphere before the veil of politics dropped over his expression: "The more deeply I search for the roots of the global environmental crisis," he said in 1992, "the more I am convinced that it is an outer manifestation of an inner crisis that is, for lack of a better word, spiritual." [18] Now that his campaign is over, the world will be a better place if Mr. Gore once again takes these words to heart and joins the green team. He is a natural ally for the coalition to form a global green republic. Another is Prince Charles, if we could trust him to challenge his powerful relatives to give up playing the global domination game.

Perhaps the most articulate spokesman for the green movement is the Norwegian academic philosopher Arne Naess, founder of the deep ecology philosophy. A portion of his Deep Ecology Platform follows:

1. *The well-being of human and nonhuman life on Earth have value in themselves. These values are independent of the nonhuman world for human purposes.*

2. *Richness and diversity of life forms contribute to the realization of these values and are also values in themselves.*

3. *Humans have no right to reduce this richness and diversity except to satisfy vital needs.*

4. *The flourishing of human life...is compatible with a significant decrease of the human population. The flourishing of nonhuman life requires such a decrease.*

5. *Present human interference with the nonhuman world is excessive, and the situation is rapidly worsening.*

6. *Policies must therefore be changed and will affect (existing) structures. The (future) will be deeply different from the present...*

7. *The ideological change is mainly that of appreciating life quality rather than adhering to an increasingly higher standard of living...*

8. *Those who subscribe to the foregoing points have an obligation directly to try to implement the necessary changes."* [8,19]

The deep ecology movement is not restricted to idealistic academics, visionary dropouts and native cultures. Many sensitive people from a variety of outlooks have made their own commitments. American author Jerry Mander sees things today as a "worldwide, interlocked, monolithic, technical, political web of unprecedented negative implications (p.4)...All these acts were and are made possible by one fundamental rationalization: that our society represents the ultimate expression of evolution, its final flowering. It is this attitude, and its corresponding belief that native societies represent an earlier form...that seem to unify all modern political perspectives: Right, Left, Capitalist, and Marxist. (p.7)" [20]

Mander sees the enormous influence that television and advertising have on the public. "Once we accept life within a technically mediated reality," he writes, "we become less aware of anything that preceded it. We have a hard time imagining life before television and cars. We do not remember a United States of mainly forests and quiet. The information that nature offers to our minds and our senses is nearly absent from our lives. If we do seek out nature, we find it fenced off in a 'park', a kind of nature zoo. It's little wonder that we find incomprehensible any societies that choose to live within nature.(p.32)...Most technologies are actually deployed in the manner that is most useful to the institutions that gain from their use; this may have nothing to do with public or planetary good.(p.74)"

"According to Advertising Age, about 75 per cent of commercial network television time is paid for by the 100 largest corporations in America...only 100 corporations get to decide what will appear on television and what will not..." (Ref. 20, p. 78)   This now includes so-called "public" television, in which more than half of the sponsorship comes from the same 100 corporations. The trillions of dollars now spent on advertising and public relations worldwide exceeds even the amount spent on education. Meanwhile, under the Bush administration, the media are being increasingly controlled by fewer corporations.

Another ecological view comes from American author William Ashworth. "We are, all of us, subject to natural law", he writes. "We can manipulate its results, but we cannot change its operation; we may be able to mask its effects on us for a while, but we will never be able to eliminate them...we look at our cities and our automobiles and our computers and our TV dinners and think we have created something. We have not. All we have done is use pre-created rules to put pre-created things in new ways."[21] Ashworth believes that even preservation efforts as well as exploitation are unrealistic because we are part of the biosphere and not artificially walled off from it. Counting ourselves out of nature in this way would be a violation of natural law. So we see many opinions within the ecological community which must be debated. Nevertheless, we must all remember that the common ground is the need to eliminate human-caused unsustainability.

In the end, most of us will have to become partial or complete converts to the deep ecology point of view in order to survive the crisis. The growing "voluntary simplicity" movement provides a model. My own conversion is an example. After a one year stint in 1975 as a special consultant to the U.S. House of Representatives subcommittee on energy and the environment, I joined the physics faculty at Princeton University to work with the late professor Gerard O'Neill on the concept of settling and industrializing space while restoring the Earth to sustainability.

Armed with new insights and experiences in the U.S. Congress on the ecological catastrophes wrought by dirty nuclear and fossil fuel energy, I was eager to comb through the renewable options. Yet I still had the desire to participate in the materialistic human adventures on vast new frontiers. Exploring space and the planets felt like the early American West to me and then some, and so captured my imagination. I then felt that a resolution of two apparently conflicting philosophies was at hand. The physics and economics argued in favor of this solution: lunar or asteroidal materials could be cheaply extracted and fabricated into industries that could provide abundant feedstock of mineral resources, food and energy to the Earth.

These space development studies seemed to reconcile my former world views of economic expansion and exploration on the one hand, with respect for the limits to growth of the Earth's resources on the other hand. We had an answer to the dilemma, I thought, a reconciliation between two paradigms which could emerge as a compromise between anthropocentric materialism and ecological sustainability.

I have since changed my thinking on this. I am opposed to the widespread industrialization of space, in fact the widespread industrialization of anything. What happened in the interim to make my conversion? First, the growing pollution of the planet since the 1970s put me "over the top" about the excesses of the Industrial Age. I began to question the morality of any form of human exploitation, even beyond the Earth's biosphere. I reasoned that we would eventually make a mess out there too. Another aspect of my conversion was the poignant experiences of having been in a variety of natural and toxic settings worldwide over decades—something we could all look at from a personal perspective, as I'll explain in the next chapter. But perhaps most important to me was a growing spiritual awareness of our own responsibility as citizens of a sustainable universe—a position from which in good conscience I could not retreat or compromise. Space industrialization turned out for me to be a double standard which would dilute our mandate to tread lightly upon the Earth. We must find a way to replace economic growth as a motive for living, or we'll all destroy ourselves and everywhere we go.

My earlier research on mining the asteroids for an economical space infrastructure has given way to the larger view of a universal ecology which respects all of nature on Earth and beyond. Maybe we could buy some time by moving polluting industry into space, but sooner or later, we shall need to make a set of more fundamental changes that would end all human pollution everywhere. Why not sooner rather than later? Why settle for just moving our garbage elsewhere? We're already doing that with radioactive waste here. I believe ultimately that we have no choice but to spiritualize rather than materialize ourselves and our surroundings.

I do not oppose open space and planetary exploration. I warmly support it if its impact is minimal and results shared freely with the public, because this activity is an important part of expanding our knowledge about the universe and our place in it. I would love to see novel, clean propulsion systems as well as terrestrial free energy consistent with a universally sustainable, ecological and peaceful ethic. In the end, the formation of a global green democracy will require widespread education, discussion and debate on what is meant by universal ecology and sustainability, at all levels of discourse. We will need to take the power away from NASA, whose new-found elitism, secrecy and militarism not only violate its public charter, but are keeping much of the truth of new knowledge away from us. [22]

To summarize, the sustainability agenda based on a deep ecology ethic such as that of Naess will need to include a simple set of powerful principles which would form a constitution for a new global democracy. We shall return to those principles in a draft of a manifesto for sustainability in Chapter 7. Meanwhile, we need to realize that our personal ecologies are becoming unsustainable too. Not only the Earth, but the very survival of the human race itself, is at stake. In fact, the collective cancer humanity represents could be removed from the Earth in ways we could never imagine. This fact alone imposes a great sense of personal responsibility for all of us to heal both ourselves and the planet. Next we will look at what we must do personally to qualify ourselves to re-inherit the Earth.

References for Chapter 4

1. Charles A. Reich, *Opposing the System*, Crown Publishers, New York, 1995.

2. Kenneth S. Davis, *FDR: Into the Storm*, 1937-1940, Random House, New York, 1993, p. 41.

3. David C. Korten, *When Corporations Rule the World*, Kumarian Press, West Hartford, Conn., 1995, *The Post-Corporate World*, ibid, 1999.

4. Alex Falconer MEP, "The Shape of Things to Come", *The Social Crediter*, Edinburgh, July-August, 1998, p. 28.

5. Alfred, Lord Tennyson, "In Memoriam A.H.H.".

6. Alexander Stille, "Ideas", *New York Times*, November 11, 2000.

7. www.foe.org, 1999 and 2000.

8. Arne Naess, "Politics and the Ecological Crisis", *Revision*, vol. 13, no. 3, 1991; also "The Three Great Movements", *The Trumpeter*, vol. 9, no. 2, 1992, pp. 85-86.

9. Benjamin B. Ferencz and Ken Keyes, Jr., *PlanetHood*, Love Line Press, Coos Bay, Oregon, 1991.

10. Paul H. Ray, "The Rise of Integral Culture", *Noetic Sciences Review*, no. 37, Spring, 1996; *Cultural Creatives*, Three Rivers Press, New York, 2000.

11. Brian O'Leary, "The Green Movement and Social Change", *Ecolonomics in Action*, Sept./Oct. 2000.

12. Peter LaVaute, "Enough for Everyone", Institute of Ecolonomics, Ridgway, Colorado, 2000.

13. Dennis Weaver, editorial, *Ecolonomics in Action*, Sept./Oct. 2000.

14. Paul R. Ehrlich and Anne H. Ehrlich, *Betrayal of Science and Reason*, Island Press, Washington, D.C., 1996.

15. Ralph Metzner, "The Emerging Ecological Worldview," in *Worldviews and Ecology*, edit. by M.E. Tucker and J.A. Grim, Orbis Books, Maryknoll, New York, 1994.

16. Mary Evelyn Tucker and John A Grim, preface to *Worldviews and Ecology, ibid.*

17. Tu Wei-Ming, "Beyond the Enlightenment Mentality", *ibid.*

18. Albert Gore, Jr., *Earth in the Balance: Ecology and the Human Spirit*, Houghton Mifflin, New York, 1992.

19. George Sessions, "Deep Ecology as Worldview", *Worldviews and Ecology*, edit. M.A. Tucker and J.A. Grim, Orbis Books, Maryknoll, New York, 1994.

20. Jerry Mander, *In the Absence of the Sacred*, Sierra Club Books, San Francisco, 1994.

21. William Ashworth, *The Left Hand of Eden*, Oregon State Univ. Press, Corvallis, Oregon, 1999.

22. Brian O'Leary, *Miracle in the Void*, Kamapua'a Press, Kihei, Hawaii, 1996.

* *My expression here about the greed and power of Enron and other large corporations and their influence on government predated the public exposition during 2002. On two occasions during 2000 I gave speeches in Houston and Montreal castigating Enron's role in bilking California's electricity consumers. I also pointed out how Enron was perfectly positioned to lead the way into a hydrogen economy: they had the natural gas pipeline infrastructure which could accomodate hydrogen, and the capital to make the national conversion. Each time, a middle manager from Enron was in the audience, and each time they felt a bit guilty and wanted to set up a time for me to make a presentation to their top management, who ended up not granting me the appointment. I guess they were too busy selling stock and managing off-shore accounts to help the planet.*

5

Clean Up and Enhance our Personal Ecologies

"We treat the Earth as if it were a business in liquidation."
-Herman Daly, economist

RETURNING POWER TO the people under a global green republic will trigger massive social change. Not returning power to the people under a global green republic will also result in massive social change—of the intolerably tyrannical and polluting variety. Whichever way we turn, each and every one of us will be profoundly affected by our ability or inability to meet the challenges before us. Either way, we shall have to grieve our old familiar ways and move into uncharted territory. If the social changes of the twentieth century were unprecedented and vast, the social changes of the twenty-first century will be explosive.

In his prophetic book *Opposing the System* [1], Charles Reich stressed that the new paradigm must include looking at the individual's needs, to tap into our own creativity, to restore our sense of vision, free of the disproportionate demands of economic government. The questions that seem to fall through the cracks of contemporary culture include: Who am I? What is the meaning of my existence, our existence? How important is it to me personally to create a sustainable future? How can I participate? Do I have any power to make a difference? How can I improve my health to become more vibrant and more aware of my choices, our choices? How can I enhance my family and community life as cleansing cells that could replace the toxic cells humans have implanted into Gaia? How can we expand our knowledge base, free of the biases and suppressions of our mainstream culture? Do we survive death? Are we alone in the universe?

This chapter and the next will focus on what an awakening individual can do about our quest for the truth, how new systems of knowledge and understanding can bring us together in deepening harmony. The fact is, we will all need to go beyond deep ecology and to learn at a more fundamental level why we are here, the nonanthropocentric significance of being a human on Earth. Our corporate, governmental, religious, media and academic leaders don't want to do that because the answers might become embarrassing to them. They espouse outmoded ways that can go back centuries. They do not want their bubbles to burst and are going for the near term profit while the sun still shines on them. But for our very survival, we will need to question their values, our own values. We must act anew.

We need a new ethics. We must ask, How are we treating ourselves and our fellow humans and life in general? How do we treat our own bodies, minds and spirits? How do we feel to be in a toxic environment? A beautiful environment? What standards of ethics can we agree upon for our individual and collective futures? The Golden Rule seems to have reverted to, "He who has the gold rules."

In the end, the answers to these questions cannot come from presidents, corporations, doctors, lawyers, teachers, professors, pundits, authors, journalists, ministers, gurus or even the growing green coalition itself. They must come from within each and every one of us. But how can we move into a deeper set of compassionate ethics if virtually every one and every thing around us is out of balance? For we are in a race in which our personal ecologies are becoming as unsustainable as those of the Earth. It is a sobering thought that we all could disappear from the face of Gaia if she chooses to shake us off her body like fleas.

We therefore need to put ourselves, our species, in a broader context. The most important conversion to a deep ecological position can happen only when we openly choose a new path, to stand up in our courage as individuals and be counted, to passionately express our position that something must be done about our social malaise. Whether by choice or by outside causes, each of us must eventually become responsible for taking part in the restoration of ourselves, our environment, our purpose and our visions.

Those of us who have made the conversion find significant changes in their personal lifestyles. This is sometimes called voluntary simplicity. Arne Naess has pointed out some individual changes common to the deep ecology movement in terms of tendencies toward: "1) using simple means; 2) anticonsumerism; 3) efforts to satisfy vital needs rather than desires; 4) going for depth and richness of experience rather than intensity; 5) attempts to live in nature and to promote community rather than society; 6) appreciation of ethnic and cultural differences; 7) a concern about the situation of the Third and Fourth Worlds and an attempt to avoid a standard of living too much higher than (that of) the needy (global solidarity of lifestyle); 8) appreciation of life-styles which are universalizable, which are not blatantly impossible to sustain without injustice toward fellow humans and other species; 9) appreciating all life forms; 10) a tendency toward vegetarianism; 11) protecting wild species in conflicts with domestic animals; 12) efforts to protect local ecosystems; and 13) acting nonviolently. "[2]

Sometimes we don't make these lifestyle changes ourselves; they are forced upon us by external events. This happened to me, when my own quest for significant economic opportunities became thwarted over a period of years. Yet in our culture it is not "cool" to admit "failure" in the material world, so my own simplification of lifestyle became a matter of quiet necessity, not a choice. In a word, I became more humble. In the process, I learned that stepping outside the economic System has its own personal rewards and challenges, and I suspect that many more of us are affected by this dynamic than we are willing to admit. So from personal experience, I would add another item to Naess' list of lifestyle changes: (14) for those of us in the West whose lifestyle changes must be significant, we new deep ecologists cannot give up our ethical responsibilities in order to return to those old comforts and securities which contradict a sustainable future; we must acknowledge the pain of our grief at the loss of the old ways and move on to the new. This does not preempt continuing to open ourselves to economic abundance, if that prosperity can be applied to the overall effort towards a green future. If the money is there, great; whether it is or it isn't available, we all can still play a significant role in a simplified and more humble form. My own story provides an example of the kind of journey we may all have to take, so that we may transcend the tyranny of these times.

Changes Imposed on My Parents and Myself

Thirty years ago was the last time I saw my father. I remember sitting at the dinner table with him and my mother, who were visiting my young family in Trumansburg, New York. That evening, my dad had had his usual stiff drinks, and just after we sat down, he coughed, choked, slumped onto the table, and lost consciousness. By the time the ambulance got him to the emergency room, he had come to and was embarrassed about the incident. The cardiograms showed nothing.

One month later my dad passed over from a coronary heart attack. Here was a classic story of a man who "died in his boots", as my mother expressed it. He was within weeks of retirement. Perhaps at some level he was afraid of what was ahead for him, as he had been focused on his work while hiding in an emotional cave. My father had been a good man, of Irish descent from Boston, who rose to an appliance sales managership. Both parents encouraged and sponsored the best education available and a comfortable upbringing.

Fortunately I left him with a hopeful farewell just before his passing, because he had been very concerned about my independent nature. I had left the astronaut program the previous year to become an assistant professor of astronomy and space science at Cornell University. In various events we had attended, my dad and I hobnobbed with the likes of Werner von Braun, Walter Cronkite, Wally Schirra and Carl Sagan. This gave him great pleasure. And I had a lucrative contract with Houghton-Mifflin to publish my first book, *The Making of An Ex-Astronaut*. Although I had resigned from the astronaut program, he was proud of me nonetheless. He saw me heading toward success in the mainstream culture, so he never had the chance to face the disappointment that his son had become a scientific heretic or dropout from the esteemed traditional circles he had worked so hard to sponsor.

My father epitomized the values of the System. When I was a school child he had often warned me I would be a shoe salesman if I didn't get all A's on my report card. He admonished me that it was economically essential to earn a substantial living to be secure and successful. I wasn't at that time aware of the demands the System would place on me, nor did I follow my father's advice — yet anyway. In a sense, he was right: opposing the System could have serious personal economic consequences. Yet I also grew to learn many things my father couldn't in his situation: I have recently had a chance to leave the System's daily demands and look more deeply into my health and my mission here.

Charles Reich believes that, in a better world, personal life needs to have recognized social value to the same degree as that of developing our material surroundings through sensible economic and public participation. "Family life, spiritual life, nature, health, trust, safety, and security are all intangible and unmeasurable but essential to survival. Unless they are assigned value in our economic system, they will be plundered..." (Ref. 1, p. 198). My life has often been a delicate balancing act between the two. I have often suffered for lack of money. I've also been able to live life to its fullest on the periphery of the culture, but getting out of balance can cause a lot of anxiety. Yet is that any more stressful than staying within the System and meeting my father's fate? My dad felt that men did not have a choice in the matter: you work for the System.

My mother, too, has recently met her fate with Alzheimer's disease. Like my father, she had had high standards and was concerned about my leaving behind a secure career as a mainstream scientist. Alzheimer's is reaching epidemic proportions in America. Half of us will have it by age eighty, according to the latest projections. It seems to afflict many of our high achievers who have worked successfully within the System. Following the example of former President Reagan, my mother's Alzheimer's home is full of former professors, doctors and business people. Perhaps the System had taken its toll on my mother as well as my father. The universe had forced both my parents to surrender to states outside the System, either through early death or brain deterioration. Their examples and those of tens of millions of others seem to be a mandate for us to avoid the System wherever possible, and to focus on our missions and our passions during our short time here in physical form. "Follow your bliss" were the late Joseph Campbell's wise words.

My parents' stories were sad because they could have relaxed and enjoyed life more. But they often worried and withdrew into their economic anxieties. My father was also in spiritual crisis, avoiding his health and fearful of what he would do to fill the void after retirement, which never came. I am often a mirror of my father. I have struggled with a tug-of-war between economic survival and my personal mission, often with my health hanging in the balance. A dramatic example follows:

A few years ago, I received a fax which was going to change my life in ways I had never imagined. I had just spent three years toiling and boiling as an author and small publisher of my eighth trade book *Miracle in the Void*. The books were now in the bookstores selling briskly and the momentum of advertising and promotion had reached their peak. I was already planning to expand my company and had often visualized achieving a hard-earned economic success. The fax came from my exclusive book distributor, one of the largest in the world. It stated that they were going out of business. In one brief moment of anguish, I foresaw my publishing efforts collapse into nothingness. With virtually all sales nullified, I scrambled feverishly along with other publishers to save my remaining inventory and to set up new distribution.

But most of the damage was irreversible and bankruptcy loomed. I soon collapsed into bed with a stress-induced pinched spinal column. The unprecedented pain did not let up for one sleepless, bedridden month. Several other survival issues such as home, relationship and further financial setbacks intensified the situation. At times I thought I was going to die or would have preferred to.

Clearly I had to do something, but answers were elusive. I was behaving erratically but always fulfilling my speaking obligations. One of these required my flying from Maui all the way to the Bahamas for only three days to give some talks on science and spirituality at the Sivananda Yoga Ashram. While I had thought it ridiculous to take such a long trip wedged between other trips, a faint voice from within advised me that I should go. And so I did.

Nestled in a five-acre tropical garden on Paradise Island, with Club Med on one side, wealthy mansions on the other, and cruise ships anchored in the harbor with Nassau in view, the Ashram was a haven of peace. The swamis, guests and students were warm, friendly and calm. Upon my tearful departure from the Ashram one of the swamis suggested, "Would you like to come to our Yoga Farm in Grass Valley, California, next month during our teachers' training course?" Without any thought, I said yes.

So for five weeks, I practiced yoga, meditated, taught new science, and mixed with the yoga teacher trainees. I left the Ashram with a brand new body and spirit, reflecting that the universe was very wise to have presented to me the challenges it did. Without them, I would never have taken up yoga for my overall health. I may not have followed my father's advice, I may not have scrambled enough to economically recover, I may not have been beholden to the demands of the System, but I acquired a new freedom to find a health insurance that no expensive health maintenance organization could have ever offered. I would never have embraced the void that needed to precede a new start. Rather than merely preaching about the merits of embracing a new science of consciousness (next chapter), I began to practice it, lest I succumb to more suffering and pain. I have learned the hard way (and, alas, perhaps only way) that we are spiritual beings awesomely interconnected with a universe which, in the end, can be compassionate and supportive if only we listen. My visits to the ashrams are teaching me a new balance personally, ecologically and spiritually.

Overcoming Economic Tyranny in our Personal Lives

In order to make the radical changes we must undergo, it seems that we cannot any longer act within the System as money-making machines. No longer should profit, power and security be our only motives for living. That is the old paradigm. The new paradigm brings us the opportunity to personally prepare ourselves for a new millennium which will not resemble the old. We will need to be tough, to endure the trials of our own impurities, to heal old patterns and to qualify ourselves to meet the enormous tasks that lie before us. As I learned in writing *Miracle in the Void*, going to the depths to grieve for life within the old System, entering the fire that tempers and purifies, can give us the courage to transform to the new—even though our new lives might be humble.

Engaging with economic tyranny can be mentally unhealthy. For example, I have experienced many occasions in which people stop being friendly when it comes to money. Going for the profit can be our greatest hidden personal agenda, based on the understandable need that some money is required for us to live. But the line between adequacy and greed is often blurred. It seems as we get closer to cash for new projects, personalities change. Promises become broken, disappointment follows, relationships chill, and the cash ends up staying where it began—within the System. This hoarding attitude is a classic psychological example of the "double-approach-avoidance-conflict", where the closer we get to our goal, the more repelled we feel by the prospect of achieving it. This hesitancy to change must be offset by a new spirit of altruism, a new ethic which places more importance on the common good than on individual economic gain.

American author Susan Ford Collins describes the competitive quest for efficiency which dominates our culture as "more-better-faster". Nowadays, says Collins, "you simply have to produce. And in an era of disposable people, if you don't keep up, someone else will." [3] She believes that we can overcome the pressure to perform by finding more creative ways to interact in the world. From this new perspective, we can truly come together to co-create our action plans. It is important not only what we do to create a sustainable future; it is just as important how we do it. We must find innovative, interactive ways of evolving our plans. The new style and content will be a lot different than those of a Bush-Cheney energy panel!

Fear of failure, fear of change, fear of never having enough, are not the only impediments to change. Many of us also don't seem to listen to our bodies in the face of economic pressure. We abuse them with environmental toxins, microbes, junk foods, smokes, drugs, alcohol, sugar, fat and televised violence. Then, when something goes wrong, we go to the doctor for pills or have a faulty part removed and/or replaced. Drug 'em or cut 'em. Then some of us become either old zombies or die young, like my father. He was sixty-seven. The good news is, more and more of us are beginning to look at health beyond the immediate hazards and symptoms, paving the way towards consciousness medicine.

Western Medicine, Public Health and Holistic Healing

When it comes to our individual health, it seems to me our future could go in any or all of at least three directions: medical business as usual, public health intervention in crisis, and the new mind-body medicine. First we have the prevailing Western model of health care. Medical professionals continue to spin expensive marvels our way with the latest high-tech diagnostic tools, surgeries and pharmaceuticals the world has ever offered. But health care has become unaffordable to an increasing proportion of the U.S. population, myself included. This has become a partisan issue politically, but there is no solution yet, as the traditional battles between "big government" and private interests continue. Economic government has co-opted medical care to the point it has become a big business rather than a birthright. Many of us in America cannot afford to see a doctor or go to the hospital. We simply must stay healthy! Again, Europe is ahead on this question. Ralph Nader and Senator Hillary Clinton have advocated guaranteed health care for everybody in America, but the System has bought out, suppressed and gridlocked any attempt to implement such a program even though it would cost less in the long run. It is clear that the civil culture is falling between the cracks—reminiscent of the issues we've already explored in energy and the environment. To qualify ourselves to re-inherit the Earth, the green team will need to move outside the System and finance its own set of preventative and interventional health care options, ranging from the source of the problem (including pollution) to its effects on the human body.

But whatever the political or economic circumstances, the familiar Western approach is clearly inadequate in the face of two cultural trends immediately upon us, showing both extraordinary danger (the toxic environment) and unprecedented opportunity (consciousness medicine and personal growth). In the end, we don't only need to become more caring towards those who cannot afford it themselves. We will need also to thoroughly address public health hazards and then step outside the box so that we may heal ourselves and Gaia. Only economic tyranny stands in the way.

I don't mean to demean Western medical practice. When my son had appendicitis years ago, we went to have it removed, the only decision that was sensible. I often regret having erred on the side of almost waiting too long during that crisis. Minutes counted. There was little question that the surgeon in the hospital environment saved his life.

But at times it's better to stay at home. My heart attack nine years ago [4] and a bad back four years ago each disabled me for weeks. This was best left away from the doctors, institutions and medication. Both ailments cost us little. Meredith's loving home care plus my later improving on my lifestyle far exceeded any healing I could have ever experienced in an emergency room, hospital room, surgery room or on drugs—even in the moment of crisis. A study of the U.S. Office of Technology Assessment concluded that "only 10-20 percent of all procedures currently used in medical practice have been shown to be efficacious by controlled trial."[5]

Using visualization techniques, I once healed a knee after refusing the doctor's advice for surgery and medication. [6] The revolution happening in the healing arts is already enhancing the health of hundreds of millions of people. According to the prestigious *Journal of the American Medical Association*, Americans made about 629 million visits to alternative health practitioners in 1997, compared with 386 million visits to primary care doctors. [7] People are spending more than $30 billion a year on complementary and holistic medicine in the United States alone. [8] Many of us have sought this approach independent of medical doctors and beyond insurance protection. The personal health revolution could become the prototype for a planetary health revolution, something we'll explore later in this chapter.

But in spite of the American Medical Association's (AMA) own admission of nonexclusivity over health care, its political clout continues with Political Action Committee donations of over $75 million to favorable candidates during the past two decades. The pharmaceutical industry alone donated $230 million to political candidates in 2000. Rather than to welcome new approaches to health care, American author Jon Rappoport writes, the AMA continues to "focus attention towards drugs and surgery and vaccines, even though the basic core of the doctor's oath is to heal by using whatever will heal." [5]

"The best analysis of the history of disease on this planet have come to the conclusion that clean water, improved sanitation, better food, higher standard of living....have been overriding factors in the decline of human disease over time. Not antibiotics, not vaccines, not other drugs, not surgery, not hospitals....Medical societies are in the business of asserting that germs are our real problem, not starvation, not contaminated drinking water in the Third World, not toxic industrial and agricultural chemicals, not any of the obvious causes of illness and death."

Recently my mother broke a hip soon after her ninetieth birthday at the beginning of a family reunion. The surgery was successful, but were it not for the fact that all three of her geographically scattered children happened to be there as her advocates during the post-surgery process, it might have been impossible for her to recover, because of the bewildering hospital environment and conflicting treatments. Like institutionalized economic government and science, institutionalized medicine is approaching a point of diminishing returns in providing us with what can truly help us. But a new health paradigm is coming on line now, because this issue is so close to home for so many of us. We can perceive more easily that our bodies are giving us the first signs of what must also happen in our environment. Both are precious and are being abused. As inside, so outside. As outside, so inside. When we heal ourselves we could also heal the planet, using analogous techniques and practices. [6] But in order to do so we are first going to have to transcend the profit-focussed medical establishment, the pharmaceutical companies, and insurance industries blocking change. It is time to provide quality health care for all on the planet under a new system which offers choices and affordable care. Some funding will have to come from the new green budgets.

"Probably we would all agree", writes Stanford engineering professor emeritus William Tiller, "that one simple prerequisite for being a good steward to a planet is to be good and conscious stewards of our own bodies." [9] I certainly agree that the same tools be applied to our physical bodies and the body of Gaia, but both activities will need to be concomitant and tireless. To re-inherit the Earth we shall need to re-inherit ourselves in a simultaneous process. We don't seem to have the time to become healthy first ourselves.

The sixty pounds of air we breathe every day, the ultraviolet radiation and bug bites that penetrate our skin, and the water and food we ingest, have more influence on our well being than we could ever realize. Climate change, ozone depletion, the foul air, the dirty and dwindling water, the contaminated food, the radioactive and chemical waste, the spread of virulent new and old strains of disease—airborne, waterborne, jetborne, and boatborne—are wreaking havoc on us.   Public health would seem to be our top priority in creating a sustainable future, at the level of direct intervention with our bodies. But sadly, our poisoned world reflects our own inability to deal effectively with the source of the problem. We're trying to make sense of all this while scrambling to re-create personal health against increasing odds. It sometimes feels as if we're rearranging the deck chairs on the Titanic.

Many people find the subject of public health difficult to address, myself included. A dark side lurks. Most of us would rather believe that the danger is exaggerated, it's happening somewhere remote and not to me, and the problem will go away. For example, at Harvard University, the medical school recently created a series of three seminars on human health and the global environment. "The first seminar", writes Ross Gelbspan, "was overflowing with energetic, aware, and concerned young medical students. Inexplicably, however, the second session was only half full. At the final seminar only half a dozen students showed up....When the perplexed (hosting professor) asked the students why attendance fell so sharply, their responses were identical. The material was compelling, they said, but it engendered overwhelming personal reactions. The problems were so great—and the ability of the students to affect them so remote—that they could deal with their feelings of frustration and helplessness and depression only by staying away. " [10]

I had a similar experience when speaking at a 1999 conference on solutions (e.g., free energy and hemp) to the global environmental crisis at my Ph.D. alma mater, the University of California at Berkeley. There was uncharacteristically great apathy there. Only a handful of people attended, mostly nonstudents. My efforts to obtain a copy of

Worldwatch's State of the World went unfulfulled as I browsed the large bookstores around the once-activist corner of Telegraph and Bancroft. Instead I saw a yuppie mall atmosphere of the campus and streets and pubs that brought me back to my undergraduate days at Williams College and hanging out in Harvard Square during the fifties. Berkeley now seemed just the opposite of what it was like when I had been there during the Free Speech Movement of the 1960s. I also discovered that the U.C. Berkeley bookstore had discontinued stocking *State of the World* as a textbook in 1995.

Our leading educational institutions join most of the rest of us who don't seem to want to hear about these things. Let someone else handle them, we think. The pendulum has swung towards apathy, yet somehow we need a middle ground of effective civil action to be able to raise and discuss and resolve the most burning global issues of our time. Berkeley was no longer in the vanguard. The System has successfully suppressed education in public health and its symbiotic relationship with the toxic environment.

What are My Experiences in Toxic and Idyllic Environments?

We don't find it easy to imagine the consequences of trees falling in Brazil, fires burning in Sumatra, or storms raging in Bangladesh. It's also not easy to sense a slowly warming atmosphere, or a brown cloud over Los Angeles that produces new lung cancer cases, or a fierce acid rain burning the eyes and lungs of people in Germany, or a bloom of algae coming from a Chinese cargo ship in a Peruvian harbor spawning a cholera outbreak, or a swarm of yellow fever or dengue fever mosquitos that manage to spread northward from Latin America to Texas, or a new outbreak of Ebola or malaria in Africa, or new cancer cases downstream from dioxin dumped from a chemical factory into a New Jersey river or downwind from a nuclear waste facility....

The fact is, we have an increasingly diseased environment, but it hasn't yet come home to most of us, especially our elected leaders and corporate managers in control. It therefore helps for each of us to reflect on our own experiences in toxic environments, how we felt, what we might begin to realize what it would be like to live in unhealthy surroundings most of the time. It's also useful to reflect on healthy environments, how life might be enhanced in a pristine or restored world.

This way, we can begin to become more sensitive to one another's health—based on direct personal experience rather than a news bite about remote suffering that we might quickly forget. What have been your most and least toxic environments and how did you feel about them?

One of my own best-and-worst environmental stories came from a single journey I took to the South Pacific in 1986 to encounter Halley's Comet and many exotic islands. I had been invited as an astronomy lecturer on board the Society Explorer en route from Easter Island to Tahiti. Easter Island is one of the most isolated spots on Earth, a five hour flight from Santiago, Chile to the East, and another five hours from Papeete, Tahiti to the West. There I met up with my roommate Bengt Danielsson, a Swedish anthropologist and author, considered to be the foremost Western expert on Polynesia. Danielsson proved to be a treasure of information about the evolution of these cultures and their sometimes devastating interactions with Europeans and Americans. He had accompanied Thor Heyerdahl on his historic Kon Tiki journey from Easter Island to Peru.

The story of Easter Island itself is fascinating. Long before the first European ship ever arrived, the once vibrant Polynesian culture had declined, probably because of the denuding of their forests, as pointed out by both Danielsson and Worldwatch. [11] Today, crisscrossing the miles of roads around the island reveals rolling treeless grasslands and rocky outcrops. The occasional coconut palms appear on the beaches alongside eerie ancient statues standing on platforms holding sentinel over ceremonial areas near the sea. Easter Island is a prototype for what the world could become.

But the real action on the trip began when we set sail westwards toward French Polynesia. Our first stop was a most exotic one. We anchored off the coral reefs surrounding the uninhabited atoll Ducie Island, hundreds of miles from anywhere else in the vast ocean. We successfully navigated into a lagoon with our zodiac boats and spent a day in this paradise, which may not have been visited for years or decades. We could tell we were a rare caller when we discovered bottles with notes washed ashore through the years. What an experience it was, walking around this one mile crescent of an island with small native bush trees holding the nests of huge frigate birds. Flora and fauna were monocultured venturers naturally blown across the vast reaches of the sea a long time ago, and are still preserved there. The pristine feeling of the experi-

ence in this spot was beyond words. The sky, sand and sea had a lumi-
nescent quality which many of us noticed. Here was a fragile and well-
defined ecosystem which had managed to escape the ravages of world
military, political and industrial interests. I stood in awe as I treaded
lightly over the sand and surf to pay homage to this simple masterpiece
crafted by Gaia. Ducie Island was the good news.

The shock of the journey struck us several days later when we
called on Mangareva. This easternmost island of French Polynesia at first
appeared as a paradise when the ship entered the lagoon outside a wide
sweeping crescent beach. At the extreme left end was an uncompleted
cathedral more than half the size of Notre Dame in Paris. At the right was
a concrete bunker built to herd the islanders inside to protect them from
the direct blasts of nearby French atmospheric nuclear tests during the
1960s.

Both structures are monuments to human folly, precursors to our
current invasion of the planet as a whole. They symbolize the harsh decimation of the population on two separate occasions of French occupa-
tion. Once a happy and stable culture of a few thousand Polynesians,
Mangareva first had to deal with an overzealous French missionary priest
during the nineteenth century who had tasked the natives to build the
cathedral. Construction finally stopped when the workers, unaccus-
tomed to hard labor, were dropping off like flies. Then the twentieth century saw a second decimation, because the nuclear fallout poisoned the
food chain right up to themselves. The natives traditionally caught and
ate their fish, and weren't about to listen to the French authorities after
they nuked the place. Many of the people we saw on that island were sick
and deformed from the toxins. These little-known stories provide a
prime example of exploitation, pollution, and negligent public health
practices.

Even the main French Polynesian paradise of Tahiti has not been
immune to the ravages of humankind. On a 1994 visit to Danielsson at
his home outside Papeete, I felt his bitterness as he talked about what was
happening to the Tahitian environment. As on Maui, most of the water
has been diverted for agriculture and tourism, drying up streams and
waterfalls. But worse, raw sewage dumped into the rivers and ocean has
made most areas on Tahiti unswimmable. In addition, the El Nino and
other climate change effects have enveloped Tahiti within the equatorial
doldrums, creating a hot, still, sticky, cloudy climate with the new addition of thunderstorms. Danielsson had lived in Tahiti for decades, but

had rarely seen such weather appear until the 1990s. Usually the climate there is fair with cooling trade winds and showers.

Does this portend an unstable future for the equally fragile Hawaiian Islands? Perhaps. Maui is facing a number of environmental issues such as water shortages, deforestation, the introduction of alien flora and fauna, growing urban and industrial sprawl, droughts, dying coral reefs, and noise and air pollution. The archaic practice of burning sugar cane produces irritating black snow and lung-congesting soot that often falls on the resort town of Kihei. The Maui Electric Company plans to build a huge dirty diesel power plant potentially costing $400 million that would make burning sugar cane a picnic for the lungs of downwind residents. And, in spite of the protests of entomologists, planners would like to extend the main airport runway that would welcome new nonstop flights from Asia and the American midwest, significantly increasing the further spread of unwanted alien species of fauna and flora onto and across Maui—following the example of Oahu and its international airport in Honolulu.

In summary, our public health is endangered no matter where we might go, including even some of the most idyllic and remote settings on the planet. And many other places I have visited have become wretched. Calcutta, Delhi, Bombay, Bangkok, Chernobyl and Beijing were sober reminders. We all have experienced at times the direct effects of polluted surroundings—the choking, the wheezing, headaches, nausea and stress. I find it useful to reflect and share with others our most positive and negative experiences.

The Coming Plague

Environmental public health professionals worldwide are confronting a hydraheaded beast as complex and ominous as the ecological situation itself. Both our inner and outer environments are inviting virulent old and new strains. In a 1959 book *Mirage of Health* which I had read as a text in college forty years ago, biologist Rene Dubos warned: "Human destiny is bound to remain a gamble, because at some unpredictable time and in some unforeseeable manner, nature will strike back."[12] At the time, I was unaware of the gravity of his prophetic statement, just as a Harvard or Berkeley student does not know today. Top priority is given instead to a career in business, law, medicine or science. Ironically, these young people may not be able to fulfill their own careers in the way they expected because of future environmental reprisals for our combined silence about what's really needed.

Any one of several disease possibilities awaits its ecological opportunity. Biologist-journalist Laurie Garrett produced hundreds of examples of how environmental deterioration and globalization greatly increase the chances of breeding and spreading of deadly diseases in her recent book *The Coming Plague*. "That humanity", she writes, "had grossly underestimated the microbes was no longer a matter of doubt. The microbes were winning...the extraordinary, rapid growth of the homo sapiens population, coupled with its voracious appetite for planetary dominance and resource consumption, had put every measurable biological and chemical system in a state of imbalance." [13]

She sums up a recent conference of concerned scientists this way: "viruses were mutating at rapid rates; seals were dying in great plagues as the researchers convened; more than 90 percent of the rabbits in Australia died in a single year following the introduction of a new virus to the land; great influenza pandemics were sweeping through the animal world; the Andromeda strain newly surfaced in Africa in the form of Ebola virus; megacities were arising in the developing world, creating niches from which 'virtually anything might arise'; rain forests were being destroyed, forcing disease-carrying animals and insects into areas of human habitation and raising the very real possibility that lethal, mysterious microbes would, for the first time, infect humanity on a large scale and imperil the survival of the human race." Garrett believes that the real situation reads more like a Michael Crichton novel than the quickly forgotten news reports, which always seem to affect others, not us.

At this writing, headlines on the AIDS situation are not encouraging. Existing treatments using powerful drugs may have temporarily lowered the death rates, but now even those are wearing off in the United States, according to an August 30, 1999 Associated Press report. The drop in AIDS deaths was halved in 1998, and AIDS continues to be an epidemic, with over 17,000 deaths that year. A cure to this ruthless killer has still not been found.

After more than a decade of investigative journalism on the AIDS question, Jon Rappoport believes that its continuing toll on human life can be traced to industrial greed. Many experts have concluded that AIDS is a general category for immune deficiency diseases rather than a particular virus searching for a vaccine. Robust drugs such as AZT actually accelerate the progression of symptoms from its early stages, based on studies of HIV-positive men in San Francisco who refused to take the drug and were able to forestall the symptoms of AIDS for years.

In an irreverent and perceptive "dream monologue", Rappoport imagines what the CEO of a large drug company might be thinking of the overall situation..."AIDS is really a collection of different forms of immune-suppression from different causes around the world. I know that the causes in many cases are chemical, or relate to horrible starvation in the Third World. But my company makes no money from that. There is no medical drug we can develop to treat that. We have to have the mirage of a single medical condition caused by a single germ. THEN we can find a drug and a vaccine and sell them. In the case of AIDS, we saw AZT come to the fore as the drug of choice, and billions of dollars of profit were made. The AIDS vaccine, if it ever comes to market, could mean profits of two or three hundred billion dollars. For that kind of money, we need to maintain our position in people's minds as the authorities in the field of disease..." [5]

Laurie Garrett's testimony of the AIDS problem was this: "Through the AIDS prism", she wrote, "it was possible for the world's public health experts to witness what they considered to be the hypocrisies, cruelties, failings, and inadequacies of humanity's sacred institutions, including its medical establishment, science, organized religion, systems of justice, the United Nations, and individual government systems of all political stripes. If HIV was our model, leading scientists concluded, humanity was in very big trouble...over the past five years, scientists—particularly in the United States and France—have voiced concern that HIV, far from representing a public health aberration, may be a sign of things to come. They warn that humanity has learned little about the preparedness and response to new microbes, despite the blatant tragedy of AIDS. And they call for recognition of the ways in which changes at the micro level of the environment of any nation can affect life at the global, macro level." [13]

The scientists in the conference Garrett attended wondered how history might judge the recent performance of the world's political and religious leaders: "Would they be seen as equivalent to the seventeenth-century clerics and aristocracy of London who fled the city, leaving the poor to suffer the bubonic plague; or would history be more compassionate, merely finding them incapable of seeing the storm until it leveled their homes?"

AIDS is but one example of a world run amok by the potential of plague. An August 30, 1999 Learning Channel documentary on diseases and their human mismanagement brought the issue into many homes. Included in this penetrating and chilling report is the ongoing threat of biological and chemical warfare options that could destroy all of us. Short of that, at any time we could encounter another influenza pandemic or Ebola-type strains that could emerge from a human-decimated rainforest, dirty river, airborne mosquito or laboratory accident. Yet somehow we tolerate existing practices while waiting for the Big One to sweep across the world, as has always happened in history. Except the conditions for robust microbes to take over the planet are more favorable than ever.

"Ultimately", Garrett concludes, "humanity will have to change its perspective on its place in Earth's ecology if the species hopes to stave off or survive the next plague. Rapid globalization of human niches requires that human beings everywhere on the planet go beyond viewing their neighborhoods, provinces, countries, or hemispheres as the sum total of their personal ecospheres. Microbes, and their vectors, recognize none of the artificial boundaries erected by human beings...In the microbal world warfare is a constant." Garrett suggests we are not necessarily at the top of the food chain, and that we may have already ensured victory for the microbes.

Reversing Our Diseased and Toxic Environment

Not only do the microbes await us. The indiscriminate dumping of chemical toxins and radioactive materials has led to an increasing number of cancer deaths worldwide. "Cancer may be lottery", writes Sandra Steingraber, "but we don't all have equal chances of 'winning'. In America alone, over 10,000 people are killed by environmentally induced cancers each year. Simply because of where they were—the land they lived on, the air they breathed, the food they ate—they died. What is crucial now is a human rights approach to the environment. Only then can we look towards a time when deliberate and routine release of known and suspected carcinogens—and their generation in the first place—is as unthinkable a practice as slavery. In short we must become carcinogen abolitionists."[14]

Following in the tradition of Rachel Carson's *Silent Spring*, Steingraber well documents the silencing of environmental and health scientists who are dependent on government funding. "Silent Spring",

Steingraber says, "can be read as an exploration of how one kind of silence breeds another, how the secrecies of government beget a weirdly quiet and lifeless world."

Steingraber goes for solutions by proposing three principles to clean up the toxic environment. The first is the precautionary principle, which states that "indication of harm, rather than proof of harm, should be the trigger for action—especially if delay may cause irreparable damage." Our current system, she says, is "tantamount to running an uncontrolled experiment using human subjects."

The second is the principle of reverse onus, in which safety rather than harm should necessitate demonstration, shifting "the burden of proof off the public and onto those who produce, import or use the substance in question..." Steingraber argues that these standards are imposed on pharmaceuticals but not yet for industrial chemicals. (And the many promotions of unsafe pharmaceuticals have their own abuse problems). The third is the principle of the least toxic alternative. This sort of "environmental impact statement" would need to address alternatives toward achieving the goal.

We saw in Chapter 1 that the most effective progress in the environmental movement nowadays in America is court confrontations regarding the health hazards of toxic waste. This kind of action is not possible with global issues such as human-induced climate change. But eventually we shall have to move beyond the adversarial process and join together in civil responsibility for the environment under a global green republic.

Rachel Carson, Laurie Garrett and Sandra Steingraber are all passionate, intelligent and literate women who endorse not only improved public health measures to avert the tide. They believe that if we treat only the symptoms rather than the causes, ill-health and self-extinction will inevitably happen. So we must therefore come back to the source of it all: what homo sapiens is doing to the environment to create this condition in the first place. Once again, we shall need to look at what we've done and how we can reverse it in order to create a sustainable future. It's not an easy task.

Yet the job must be done. The public health issue is the most immediate effect of the declining environment itself, and needs to be supported by the green agenda. Scientists everywhere are telling us we are in harm's way of human-wrought death and disease that would make any war we have fought appear to be tame. This should give ample warning to follow the principles of Part I about restoring the biosphere, and hasten the plans we as a global community will need to follow soon.

In vivid contrast to environmentally induced public health challenges, we will also want to look at consciousness medicine, whose miraculous potential could accelerate the prospects of empowering ourselves both as universal humans and stewards of a restorable Earth. We must change our own inner paradigms as we begin to see it in the outer. Our task begins with a deeper understanding of the human body, mind and spirit as revealed by a plethora of traditional and new medical practices which transcend the Western model based on materialism and economics.

The Consciousness Medicine Revolution

Earlier in this chapter we looked at the profound lifestyle changes we must make as deep ecologists. This step will become necessary for our own survival. The six billion of us here, particularly those of us who consume more than our fair share, must learn to be personally sustainable, to simplify, to re-examine what's really important in terms of our new awareness of our own individual responsibility. Along with our outer green behavior we shall need to develop an inner green ecology, one which embraces an attitude of positive health in contrast to the barrage of commercial promotions by the pharmaceutical industry and health maintenance organizations. We must learn that we are more in essence than a machine with parts that are fixed as if by a garage mechanic. Fortunately, we have many choices for deepening and enhancing our personal ecologies than we are told by doctors and the media. For we ourselves are vehicles of consciousness awaiting self-discovery.

This coming trend in medicine has many different names, many different modalities. Taken together they comprise a revolution whose understanding has barely touched the radar screens of the System. Like the the cultural creative movement, the consciousness medicine movement is not yet organized enough to have a unified front which would be understood by an AMA or the "legal" drug cartel. Here are some of the many approaches: mind-body medicine, energy medicine, vibrational medicine, homeopathy, alternative medicine, integrative medicine, holistic healing, East-West medicine, ayruvedic medicine, Chinese medicine, acupuncture, acupressure, herbal medicine, chiropractic, and spiritual healing. Many pioneering individuals promoting these practices include

Deepak Chopra, Larry Dossey, Bernard Siegel, Andrew Weil, Elmer Green, Norman Sheely, and others too numerous to mention here.

As we shall see in the next chapter, the common denominator of most of these approaches is consciousness, which represents our ability to heal by a positive mental attitude. This new multibillion dollar industry shows clearly that these approaches both work and are popular, especially in preventing serious problems that could later arise. Perhaps most importantly, consciousness medicine opens the door to consciousness science, which, as we will see in the next chapter, will transform our paradigm to elegant ways to bring the human condition back into balance with the natural world. The box of Western medicine and science is opening now and it is a Pandora's Box for the materialists still hanging on to old ways. In the long run, scientific reality cannot be legislated or dictated by economic government, materialistic science or religion.

But we also need to look at providing alternative opportunities to those who cannot afford the kind of health care we all deserve. We need to preserve and provide natural medicines and healthy vegetarian diets, even if partial, to those of us who are ready for them. Both can come only from a biodiverse environment, mostly in our dwindling rain forests. All this argues even more strongly for preserving our natural heritage for as long as humans can do something about it. Unfortunately, holistic health care, natural medications and organic produce are beyond the financial reach of most of us, and this fact argues for universal medical insurance in which the individual can chose his or her own modality, food and medicine. Opening these choices would be a task for the global green republic.

My extensive scientific research over the past two decades supports the credibility of consciousness medicine, healing by prayer, and mind-over-matter experiments which have been ignored by the mainstream. Some religious and scientific belief systems have also been shattered by two other significant developments in understanding who we are and why we are here: mounting evidence for the survival of death, and the presence of other intelligent life in the universe. Both developments are open to scientific inquiry using the scientific method, but are culturally unacceptable because they do not fit within orthodox science, even though most religions believe in an afterlife. The System wants to retain this curious duality, another example of its tyranny. Its attempts to box in knowledge through the prohibition of looking into such fundamental questions may have culturally succeeded within universities, governmental institutions, corporations, news rooms and foundations. But this is a big lie: the scientific evidence is compelling that we do have souls, that we do carry on after what we call death, and that we are not alone in the universe.

Research Shows that We Survive Death

The positive results of remote healing and remote viewing experiments, millions of reports of out-of-body experiences associated with death, near death, ketamine and mystical encounters, verifiable past-life recall by children, and communication with deceased loved ones, all point to a nonmaterial, nonlocal and enduring reality accessible to all of us regardless of our beliefs. In spite of the materialistic cultural party line uttered by scientists and philosophers alike, with overlays by organized religion, I grew to discover the richness of empirical evidence and my own experience that we do indeed have souls which endure the demise of the physical body which we call death. [4,15] That was good news to me and should be good news for most of us, because the likelihood that our consciousness survives death gives us a deeper meaning to our existence and our responsibilities during our stay here on Earth.

I am reminded of the story of the rich merchant who gathered all his children around his death bed. His last words were, "Who's minding the store?" I often wonder what might have happened to the man just as he was making his transition. He probably would have been surprised that mundane matters faded rapidly from his consciousness, based on case studies of near-death experiences. Most of us report a feeling of oneness and a compassion for all creation—a far cry from economic tyranny.

Before I began to research these subjects I unwittingly plunged into these personally uncharted waters on a few occasions. In 1982, I was in a high-speed automobile accident which I was lucky to survive. I had sensed none of the violence of the accident, but instead felt my awareness lift out of my body towards a brilliant and timeless light from which I did not want to separate. Only later did I learn that millions of others have reported similar experiences upon their encounters with death. Some years later, I was in a clinical experiment in which I took the mind-altering drug ketamine. Again, my awareness left my body, I journeyed into richly visual mystical realities in which I became poignantly aware that my personal Earthly concerns paled before my newly expanded awareness. Again, I later learned I was not alone, that thousands of others have had similar journeys. [6]

Science itself and my own experience combined to provide the very evidence I needed—and we shall need—to open humanity to this new paradigm of reality. The ever- growing evidence for life after death and for transcendental experience contradicts the assumptions of our prevailing cultural religion, scientism. These self-styled scientistic "skeptics" of the twentieth century have not experienced and therefore deny the existence of a nonmaterialistic, nonlocal and unitive reality. They have unnecessarily limited themselves. For the truth to be known, we must therefore question this view just as we question dominant corporate polluting practices. I slowly learned that human nature is not what the modern Western culture was telling us: in essence, we are not our bodies, and that our awareness spreads beyond time and space. Science itself was revealing what some religions and mystical traditions were saying all along—not what our influential materialistic academics continue to assert through their media mouthpieces.

These experiences are available to all of us. In his later years, the great psychologist Carl Jung lamented the Western philosophical ban on believing in our own immortality, because that was a matter for the religionists. He called the existence of the human soul as "the buried treasure in the field" of scientific inquiry. [4] After years of my own surveys on near death experiences, reincarnation and mediumship, I am convinced that the human story is more far-reaching than that assumed by the mainstream culture—both secular and religious. The arbitrary separation of church and state might be politically wise, but everybody suffers when the knowledge both groups promulgate, represents a kind of closed-mindedness and intellectual dishonesty, because the facts could threaten their vested powers. In no way should these biases prevent us from telling the greater story of our being, based on irrefutable scientific evidence.

In this respect, we live in a time similar to that of the Italian Renaissance. The prevailing paradigm of the all-powerful Catholic Church then was geocentrism (the belief that the Earth and all of humanity were at the center of the universe). Now it is anthropocentric materialism (the belief that humans dominate the Earth and are entitled to exploit whatever they'd like, following some arbitrary rules of what we call the economy). During the Renaissance, it took the scientific observations of Galileo, Bruno, Copernicus and others to unseat the geocentric paradigm. Now it might take the evidence of non-local realities, the survival of death and the presence of nonhuman visitors to unseat ourselves from our materialistic anthropocentrism. I believe that in order for our culture to survive, we shall have to transcend this prevailing worldview. We need to replace it with a universal deep ecology in which humans take their proper role in a complex hierarchy of beings; we are no greater and no less than others.

The stakes become even higher when we contemplate the evidence coming from archetypal near-death, death, and after-death experiences. At the threshold of death, many experiencers report a life review in which their deeds are morally assessed. Communications with "the other side" reveal a new life-after-death which is more perfect than life on Earth. The story of our existence here then takes on a new ethical meaning, consistent also with religious and scientific traditions of the East. According to this view, we humans incarnate upon the Earth in order to gain experience and to grow. At the time of physical death, our lives are reviewed. Our souls then can acquire a larger context for knowledge about ourselves and the universe. Later our souls might reincarnate into another human body for more learning. Some philosophers call this process karma, a universal law of cause and effect in which our individual human deeds, good and bad, will be later reflected upon us by future events either in this lifetime or a later one.

Extensive scientific evidence and personal experience back this hypothesis of reality [4,6,16]. If all this is so, we can place ever more importance on the effects of our deeds. In this view, our materialistic anthropocentrism can create not only a mess "out there" in the environment; it can also create a mess within our own everlasting psyches. What we sow we must reap. One additional thread of evidence appears to confirm this view: our encounters with nonhuman intelligence.

We are Probably Not Alone in the Universe

It should be no surprise that the hypothesis that nonhuman visitors are here has both strong scientific support from researchers and denials from the cultural mainstream. Again, our anthropocentrism puts this inquiry out of our academic boxes because the implications contradict our desire to be "king of the mountain" here on Earth, thus spawning a pervasive cult of governmental secrecy. Vested powers in control are too threatened by the knowledge that our human existence may be very humble in this small corner of the universe.

Yet the evidence for cosmic visitors is overwhelming. Years of research others and I have done on encounters with UFOs, aliens, angels, nature spirits and perhaps Gaia herself reveal not only a transcendent reality [6, 16], but a quality of deep concern about what we are doing to our precious environment. [17] In clinical studies of patients who claim to have been abducted by non-humans, John Mack, professor of psychiatry at Harvard University, has found an ecological common denominator to their experiences. Mack writes: "I was astonished to discover that, in case after case, powerful messages about the human threat to the Earth's ecology were being conveyed to the experiencers in vivid, unmistakable words and images. The impact of these communications is often profound and may inspire the experiencer to work actively on behalf of the planet's life. Indeed, it seems to me quite possible that the protection of the Earth's life is at the heart of the abduction phenomenon (pp.87-88)...Abductees are repeatedly informed that the hybrid (breeding) project is related to the perilous state of the Earth's ecology and is being conducted for the purpose of both the human and alien species (p.123)...(the Earth) is indeed a jewel in the cosmic crown, and our increasing destruction of its life appears to be a crime of cosmic proportions." (Ref. 17, p. 275).

The abductors show graphic images and "take" the experiencers to earthly environments that "demonstrate one or another aspect of the Earth's beauty and threatened ecology. Scenes of apocalyptic destruction may be juxtaposed with images of beauty so exquisite that it seems at times as if a cosmic teacher were trying to reach the experiencers in the depths of their soul." (Ref. 17, p.15) We can learn a lesson not only on reversing the extraordinary impact of our destruction, but on the method by which we must awaken to the danger. My own experience in beautiful and toxic environments described earlier in this chapter had a profound impact on me, and suggests that we all need to become more aware of our surroundings and that they might be gone some day because of our carelessness. Quoting grim statistics from *State of the World* is one thing, but converts to deep ecology may also need to directly experience the loss we would inevitably suffer if we don't change our ways.

The evidence for the presence of extraterrestrial and interdimensional visitors supports the idea of a living, interconnected and conscious universe. Mack's clinical studies show clearly that we may have no choice but to become cosmic deep ecologists. When we see all creatures and Gaia herself as friends deserving of their part in the overall scheme of things, we must inevitably change our worldview. From all indications, we must co-create with Gaia a combined future—if she'll let us after our abysmal record. It is ecocide to smugly dominate and destroy her.

The Official Cult of Secrecy

The UFO/alien question opens a Pandora's Box of new knowledge which seems to reflect more on the continuing efforts of the U.S. government and others to suppress the truth than on the high quality of outside investigations showing the phenomenon to be very real. Attempts to bring the facts of these encounters to public light are continually thwarted to the point that we often feel bludgeoned into turning away from the truth, that we should expect nothing more from our elected officials. [18] So instead of holding them accountable, we have wallowed in a box of accepted knowledge which is incomplete or untrue. We have grown to believe that opening Pandora's Box would so alter our cultural worldview this would surely destroy our existing sense of societal security. I say we shall need to open that box, regardless of the consequences. The truth shall set us free.

Unfortunately, the trends in America seem to be in the opposite direction. With the late 2000 passage of the Official Secrets Act, and without any public debate, the U.S. Congress has given the executive branch even greater powers to prosecute anyone who discloses any information it deems to be classified, under a system which is highly arbitrary in the first place. Now any American president's administration could have a field day suppressing any information it wants while telling lies and prosecuting whistle blowers. This act virtually ensures a continuing policy of UFO nondisclosure and other forms of secrecy, as is the 2001-02 Patriot Act. Repression in America is on the increase, I'm sad to report.

I also believe that our allowing these secrets to prevail is but a part of allowing a pervasive cult of secrecy to rule us in an insidiously totalitarian way. This represents nothing less than our selling out. We give faceless individuals the power to keep the embarrassing truth away from us in exchange for propping up the economy, which we are told is essential for us to continue our consuming lifestyles, but which is actually killing us. This is at best a shaky deal and at worst the end of human civilization. The secrecy further begs the question, who is behind our greedy ways, the anthropocentric grip humans seem to have on the natural world? Is there a nefarious master plan of which we are unaware? The Disclosure Project described in the Introduction is making significant progress in revealing the hidden agenda of UFO and free energy cover-ups.

I've never been a conspiracy theorist, but I have been willing to entertain evidence which is consistent with a conscious conspiracy among a hypothesized secret controlling elite to dominate the world. The evidence is there. For example, Jon Rappoport believes that the world is run by seven powerful international cartels: government, military, intelligence agencies, energy, money, medical and media. David Icke, a British journalist and author, gives evidence that the ruling elite may even be dominated by extraterrestrial groups along with an entourage of humans with secret plans to rule the planet. [19]

The evidence for these ideas is not as clearly corroborated as, say, the evidence for our wanton ecocide itself, or evidence that capital accumulation driven by greed leads to unwise decisions, or evidence for the cult of secrecy blocking our path to understanding the truth of our greater being, or evidence for the cosmic relevance of destructive human activity on Earth. But I cannot eliminate the possibility of a premeditated conspiracy because the facts do square with the theory. Whether or not the conspiracy is carefully planned, one thing is clear: it does exist, and it thrives on our silence of doing anything about it. In accepting our cultural status quo, we have in a sense become co-conspirators. In opposing our destructive ways and our secrecy, we can lift out of the conspiracy. Armed with the truth, we can then translate our protests into effective action. Abraham Lincoln made that point very clearly at the beginning of Chapter 4. We must become spiritual activists.

The issue of what's going on at deeper levels can become very complex. Not only do we see forces which seem to want to dominate and destroy the natural ecology of the Earth, but those which could be protecting us as well. I often wonder how we could have gotten this far without nuclear or chemical-biological wars. Perhaps we have overseers who will not allow that to happen and so sabotage any efforts toward such mass destruction. There have been close calls in our nuclear history. Also it should come as no surprise that ET visitations often occur at nuclear and other military facilities. Of course, all this is speculation, yet worth pondering. We need all the help we can get to create a sustainable future, and we don't seem to have the discipline to get rid of weapons of mass destruction. The Bush administration's Nuclear Missile Defense and space weapons programs merely escalate the arms race to new destabilizing levels.

I fear I may have left some of you behind in this chapter. I have made at least some controversial points: 1) sooner or later, we will need to reject economic tyranny and become a deep ecologists, simplifying our lifestyles and becoming members of sustainable   society; 2) our toxic environment could kill us all, without adequate public health measures and alternative healing practices that go beyond materialistic Western medicine; 3) our consciousness probably transcends physical death, and we might eventually confront our moral duty to tread lightly upon the Earth and the universe; 4) nonhuman visitors to Earth are dramatically telling many of us to change our ways or we will face certain ecological disaster; and 5) an official cult of secrecy diverts this information away from public awareness in order to perpetuate the status quo; and 6) we will eventually awaken to these truths. It is conceivable some people could also face karmic reprisal for excesses they have created. Perhaps they (we?) are repeating similar mistakes to those made some time ago by the visitors in their own environments. They warn us we are doomed to repeat these mistakes unless we change our course.

I am not asking you to believe all these things. But the evidence clearly points to their reality and should put icing on the cake of our need to radically change our worldview. As soon as we can open ourselves to new knowledge bearing on the broader questions of the human story, we will be able to enter a new paradigm of science which embraces these studies. As we shall see in the next chapter, this worldview of consciousness, of our interconnectedness, goes far beyond our bankrupt anthropocentrism, and provides a basis for a new universal deep ecology—one which could be enduring and return Eden to Earth. These are some of the reasons why we must passionately oppose the prevailing worldview at every turn, and replace it with something new. I believe that for us to survive we must all become deep ecologists with our every breath, before it's too late.

El Nino and our Children

Our children and grandchildren will be representing humanity in the quest to re-inherit the Earth. At the moment, we are bequeathing to them a mess. How soon can we turn it all around? I am reminded of the story of another child: El Nino. This is a climate phenomenon noted by Latin Americans one Christmas (El Nino was named after the Christ Child). The El Nino is a hot spot in the South Pacific which occasionally moves eastwards, spawning destructive weather patterns all over the world for months at a time. Intense storms become common during an El Nino, and local climates can go topsy-turvy. Cold places can become warm and warm places can become cold. Wet places can get dry and dry places can get wet. For example, one of the dryest spots on Earth, the Atacama Desert in Chile, can receive torrential El Nino rains.

Climatologists believe that the unusually strong El Ninos of the 1980s and 1990s are caused by global warming through carbon dioxide emissions. [10] El Ninos are powerful heat engines that drive the Earth's climate. Like the warming of the polar regions, these phenomena are the barometers, the thermometers, the canaries in the cage, for measuring the fever of Earth. They are the children of Gaia breaking into migrating hot and cold flashes with increasing frequency, robustness and destructiveness. Could these events be the consciousness of Earth speaking back to us?

Fevers can cleanse and purify an illness. Could we ride out Gaia's fever, nurture her and heal her pain? Could we unify in consciousness to create a magical transformation to greater wisdom, greater existence? Could we give this gift to our children? I believe the answer is yes, but the tools can only come from a unified effort among humans and nonhumans. These ideas will be described in the next chapter.

References for Chapter 5

1. Charles A. Reich, *Opposing the System*, Crown Publishers, New York, 1995.

2. Arne Naess, "Deep Ecology and Lifestyle," in *The Paradox of Environmentalism*, edit. Neil Everndon, Symposium Proceedings, Faculty of Environmental Studies, York University, Ontario, 1984, pp. 57-60; see also George Sessions, "Deep Ecology as Worldview" in *Worldviews and Ecology*, edit. M.E. Tucker and J.A. Grim, Orbis Books, Maryknoll, New York, 1994.

3. Susan Ford Collins, *The Joy of Success*, HarperCollins, New York, 2003.

4. Brian O'Leary, *The Second Coming of Science*, North Atlantic Books, Berkeley, Calif., 1993.

5. Jon Rappoport, *Notes on Scandals, Conspiracies and Coverups*, Truth Seeker, SanDiego, 1999, pp. 13, 15, and 17. He quotes a U.S. OTA report as Commerce Document, PB-286929, September 1978.

6. Brian O'Leary, *Exploring Inner and Outer Space*, North Atlantic Books, Berkeley, 1989.

7. *Journal of the American Medical Association (JAMA)*, November 11, 1998.

8. *Time* magazine, June 24, 1996.

9. William A. Tiller, *Science and Human Transformation*, Pavoir Publications, Walnut Creek, CA, 1997.

10. Ross Gelbspan, *The Heat is On*, Perseus Books, Reading, Mass., 1997.

11. Lester R. Brown and Christopher Flavin, "A New Economy for a New Century", *State of the World*, Worldwatch Institute, Norton, New York, 1999.

12. Rene Dubos, *Mirage of Health*, 1959.

13. Laurie Garrett, *The Coming Plague,* Farrar, Straus and Giroux, New York, 1994.

14. Sandra Steingraber, *Living Downstream*, Addison-Wesley Publishing Co., Inc., New York, 1997 and Virago Press, London, 1998.

15 Charles Tart, edit. *Body, Mind and Spirit,* Hampton Roads, Charlottesville, Virginia, 1997.

16. Brian O'Leary, *Miracle in the Void*, Kamapua'a Press, Kihei, Hawaii, 1996.

17. John E. Mack, M.D., *Passport to the Cosmos*, Crown Publishers, New York, 1999.

18. Steven M. Greer, M.D., *Extraterrestrial Contact: The Evidence and Implications*, Crossing Points, Inc., Afton, Virginia, 1999.

19. David Icke, *The Truth Shall Set You Free*, Bridge of Love Press, Isle of Wight, 1997.

6

Create and Support a
New Science of Consciousness

"No problem can be solved from the same consciousness that created it."
-Albert Einstein

I WRITE THIS in Hana, Maui, Hawaii. Another squall is coming in from the ocean, starting the buildup of tropical clouds at the edge of a hurricane which passes to the south. The frequent squalls are mana from nature at her best. Clean air blows in from thousands of miles of ocean by trade winds dumping cleansing showers from darkening clouds that climb the steep mountainside jungles behind us. The locals' resistance to development has preserved this jewel of a paradise, and global climate has restabilized long enough for the trade showers to resume.

We had lived in Hana in early 1998 during an El Nino drought, which had uncharacteristically parched the land. Waterfalls had dried, as we walked along the evaporated flumes and pools. I was to have written this book then, but somehow never found the inspiration; also global events such as El Nino had always seemed to overtake my expression. Perhaps I too had become parched, replacing productivity with an intense curiosity about the El Nino phenomenon itself, which was spinning cyclone storms well to the north of us, one every week, across the Pacific all the way from Japan to California, ruthlessly slamming into the U.S. West Coast, then spewing killer tornadoes in the American South and ice storms in Quebec, and causing damage of historic proportion. It had been during this time of personal and environmental drought when my studies reached the point when the accelerating global human intervention overwhelmed any public discussion of sensible solutions such as clean and renewable energy development.

Now things are wetter in Hana. Last night's twilight shower caught us by surprise while we were swimming in a nearby natural pool. This treasure of a spot not known to the tourists is a lava-ringed, clear, deep pond almost the size of a football field. It is next to the ocean barricaded from it by a bar of loose lava rock. The swim felt magical. Although we've had plenty of rain in Hana, the other side of Haleakala volcano is in drought. The residents and farmers of upcountry Kula who supply the island with an abundance of organic vegetables have had to go on water rationing while the big businesses of sugar cane, golf courses, cattle farming, condominiums and resort hotels stay green with unrationed water.

Much of that water comes from the East Maui watershed above us, rising two vertical miles through an exotic rain forest. A beautiful waterfall used to drop into the pool before it emptied into the ocean. Now the waterfall and the streambed above it are ghostly dry except for a few billibong potholes that breed mosquitoes. The water level in the pool is almost down to its 1998 level, replenished only by occasional splashes of sea surf during rough high tides, like now. It has become salty, still and warm.

The water upstream supplies the parched other side for the tourists and agribusiness. We hope for the return of the stream next winter. I reflect on the enormity of our restoration task, at how many human institutions will have to change, about how staggering (but still solvable) the problems we have created are, as expressed in these writings. In and of itself, our treatment of the biosphere can cause deep depression about our seeming lack of will to move towards sustainability.

But a kind of magic lurks around the corner—something radically new. I believe that some unexpected results of experiments coming from all traditional branches of science will change things enormously. These anomalies seem to make the materialistic world view feel silly and outmoded. Patterns are emerging which will dramatically alter how we think about ourselves, our relationship to space and time, the power of our intention over the physical universe, our interconnectedness and our eternal being. It also seems obvious that if we can use consciousness to heal ourselves, we can use it to heal the Earth. This chapter describes experiments pointing to extraordinary opportunities we now have to clean up our habitat and to expand our awareness beyond the material-istic paradigm.

I now share some material which is optimistic and easier for me to address. It has been my life's work to address the question, how can the methods of science assist us in getting out of the mess we created, and redirect us to our higher purpose of being here? In spite of obstacles I see great hope.

Today's Frontier Science Becomes Tomorrow's Reality in Spite of Appearances

Whether we like it or not, science and technology still lead the way toward cultural change. I reflect that so much of what we notice in this precious ecosystem of East Maui still comes from the twentieth century. The power poles, wires, automobiles, helicopters, wood chewers, power saws and dry river beds are all signs of the strange onslaught of humanity during this past century. Most other inhabited places on the planet are saturated with it all and cut off from nature. Almost all of these alterations of the landscape came from the minds of visionary sci-entists in the late nineteenth and early twentieth century. Yet little has changed in some of our basic thinking about the way we use energy, food, water and shelter.

We're not only living within technologies developed more than a century ago; we're also steeped in a debate of the nineteenth century sci-ence of Darwinian evolution versus the creationism of the Bible. As of 1999, some schools in Kansas didn't even teach evolution. But, oddly enough, the debate is also opening new possibilities. Recently, some sci-entists are challenging those aspects of the theory of evolution which dic-tate that a competitive "random selection" and "survival of the fittest" must apply. Neo-Darwinism is too materialistic and simplistic to fit the real picture. Rather than invoking a literal Biblical creationism, these sci-entists critical of the orthodox theory of evolution posit that there must be an "intelligent design" within evolving biological systems. This approach represents, to me, a sensible alternative that acknowledges the sheer wonder of an evolving natural system, and the intelligence behind its orderliness. Attempts by humans to manipulate ecosystems provide no match for intelligent design. In fact, the intelligent design itself may be the act of a higher consciousness. More on this later.

Meanwhile, most of us are still using old science and religion in our education and for many of our grossest material creations: internal combustion engines, trucks, cars, buses, trains, drilling rigs, refineries, urban sprawl, highways, boxy buildings, factories, boilers, grids, strip mines, dumps and flying machines. In 1900 nobody could have begun to imagine the extent to which our lifestyles would change in fifty to 100 years because of the mass production of technologies devised then. It follows that research now in the frontier sciences might set the pace for the new century, for the better I hope.

I am hopeful in part because new science is my forte, and I know of many developments on the horizon that will not only help clean up the Earth but will provide us with the tools for personal growth and awareness, our origins, our place in the universe, our greater essence, and immortal nature. Since the separation of church and state about 300 years ago, we have been increasingly fragmented about understanding who we are and what we could become. Fortunately, science itself is telling us clearly that neither materialistic atheism nor orthodox religion are all there is to reality.

For more than twenty years, my main purpose has been to investigate research in the frontier sciences. These include experiments and demonstrations which confirm anomalies that cannot be readily explained by traditional scientific theories and other assumptions about our reality. This line of inquiry has sent me in directions I would have never dreamt of as a mainstream physicist at Princeton University during the late 1970s. It is a continuing adventure which has proven to be both rewarding and sometimes downright frustrating.

Three personal experiences propelled me in this direction. The first was an ability to "tune into" or remote view a person. The second was an anomalous healing of "severe advanced arthritis" in my knee primarily by laying my hands on my knee and visualizing light coming into it. The third was a near-death experience during an automobile accident, in which I became "at one" with a bright light from which I didn't want to return. I was amazed to survive the crash, let alone be uninjured. [1]

In my quest for explanations of these events, I discovered that not only are phenomena outside the box of science open to our experience—they are also often open to scientific inquiry using rigorous scientific methods. While I found that many charlatans can make exaggerated claims, several others behind the scenes are getting true anomalies that demand new perspectives about our reality. I travelled the world many times over while meeting with some of the best and brightest researchers, doing experiments with them, and studying their (often peer-reviewed) publications in a variety of fields. I also visited and confirmed evidence of many "miracles" of materializations by the Indian swami Sai Baba and the Brazilian psychic Thomaz Green Morton. Their well-documented antics are difficult if not impossible to explain away as sleight-of-hand. The result of my unusual path is contained in three recent books comprising my New Science Trilogy. [1-3]

Observations such as the observer effect in quantum mechanics, psychokinesis, remote viewing, anomalous healing, UFOs, abductions, crop circles, precognition, near- death experiences, reincarnation, mediumship, free energy effects have all been investigated scientifically and verified as anomalous. It only takes time to integrate the widely scattered data, but once this is done, some patterns begin to emerge. From all this I proposed that we need a bigger box of scientific inquiry to embrace anomalous phenomena—a new science. Among many of these experiments, I found a common denominator some of us call consciousness. We'll soon be turning to that.

What I hadn't anticipated in fulfilling my purpose was the outright hostility that has been leveled towards many of my new colleagues and myself. Case after case came to my attention about the irrational suppression of new science at almost every turn. Like the Catholic Church in Galileo's time, now mainstream science, government, corporate interests and the media are entrenched to keep the status quo going. New science has become heretical, creating the forces of ignorance, complacency, fragmentation and polarization that compare to those in Galileo's time. Scientists nowadays have become so specialized and self-limiting within the System, most of us can't see the forest for the trees.

In Chapter 2 we looked at the example of cold fusion, where a very significant discovery has been often repeated over the course of a decade, yet continues to be suppressed and ridiculed by mainstream nuclear physicists, the U.S. Patent Office, the U.S. Department of Energy, the corporate world and the press. Individuals are warned not to do any of this research lest their funding get cut off and careers end. I slowly discovered that all science outside the box was in for tough sledding in this culture. [1-4] I will be giving several examples of promising but suppressed new science in this chapter.

Scientists themselves can behave irrationally when it comes to new paradigms. They deny new data, reject new hypotheses that contradict existing theories and punish their own colleagues who dare to explore. One doesn't require a massive CIA conspiracy or coverup to suppress science, because institutionalized scientists themselves are very good at doing that. Peer pressure and funding keep them within the System. When it comes to science outside the box, scientists are about as unscientific as one could get. [2]

Biophysicist Beverly Rubik cites a study in which a testable but unexpectedly false hypothesis was presented to a group of scientists and a group of clergymen. Neither knew that the hypothesis (which is basically a scientifically educated guess about reality) was false. "The results", writes Rubik, "showed that most of the scientists refused to declare the hypothesis false, clinging to it longer despite the lack of evidence. The clergymen, however, more frequently recognized that the hypothesis was false. This and other studies show that scientists are at least as dogmatic, authoritarian, and irrational as nonscientists in resisting unexpected findings....Most scientists prefer to elaborate what they think they know rather than focus on what they don't know."[4], (p.30).

Rubik and I have learned from decades of direct experience how mainstream science has become like a priesthood. Appendix III gives the results of a poll in which the general public was eleven times more likely to accept the possibility that psychic phenomena are real than members of the prestigious U.S. Academy of Sciences. In this case, the most elite of scientists largely refute a hypothesis which has proven to be true, based on overwhelming evidence.

The System has decreed where the boundaries of the box are to be, and prevents inquiry outside it at funded institutions.   Those who dare venture outside feel they have to be very cautious and have therefore focused their attention on a particular area only, lest their already lost credibility erode even further. For example, a UFO researcher might be quite skeptical about the existence of anomalies on Mars, a cold fusion scientist might doubt the existence of UFOs, a parapsychologist might question the reality of cold fusion, and a quantum physicist might doubt zero point energy and paranormal phenomena. My attempts to embrace these and other subjects has often removed me from the favor of some specialists in the new science . While it's certain that old paradigm scientists have given up on me as a lost cause, many new scientists also question my credibility because I accept inquiry into so many strange topics. I often feel isolated, having been invited to speak at universities only rarely and even those are sometimes rescinded or regretted (Appendix III).

But I have new colleagues scattered around the globe. A part of me would still like to be at a university because of its support structures for doing research and the pleasure of interacting daily with bright young people and peers. But in balance, my life is better spent in working with kindred spirits in organizations such as the International Association for New Science, the Institute of Noetic Sciences, the Scientific and Medical Network, and the Society for Scientific Exploration (SSE), the University of Philosophical Research, and the Sivananda Yoga Ashrams. Yet even new science organizations can sometimes act elistist. For example, in Appendix III, I present some material on an invited talk I gave in June 1999 at the annual meeting of the SSE on the suppression of science. Ironically, because of the biases of some council members, I was uninvited for awhile; a talk on suppression was nearly suppressed. As a long-standing SSE member, I protested and they reinstated the invitation. More recently, they demoted me from member to associate.

In truth, scientists—even new scientists—can be very opinionated about other frontiers without really knowing what is going on.   It doesn't seem to make any difference how competent a scientist is in his or her own field. What makes matters even more confusing is that the majority of nonscientists who accept the new data can also accept esoteric concepts, often presented as science. New scientists like me are frequently lumped with groups dealing with those purveying pseudoscientific material as well, even further subtracting from the credibility of true scientific inquiry outside the box. There are exceptions such as David Lorimer at the Scientific and Medical Network in Britain. He uses the phrase "openness with rigor" to describe how the scientific method can be applied to many basic inquiries. The process of science demands a formal set of experiments whose results withstand the tests of repeatability and replication by others.

So it's no wonder that the waters are muddy, that the visions of new science are available to few, as we struggle to build new models of our reality to fit the data. The fragmentation and polarization of specialized groups make it very difficult to share a bigger picture, even with kindred colleagues in a given field. We are going to have to cross disciplinary boundaries if we are to make any progress in defining the emerging paradigm. Political and business leaders, the media, and public seem to have little understanding of these dynamics. Yet in spite of our heavy censorship of new science at the millennial boundary, time is on our side because of the growing robustness of the data and the need for solutions to our environmental dilemma, such as free energy and a global green republic.

About one hundred years ago, electricity, telephones, cars, airplanes, radio communications and oil extraction were poorly understood. If history is a lesson, the unsung work of now will set the pace for the new century. Tomorrow's new energy devices will inevitably replace today's internal combustion engines and power systems. Today's cars and planes will become obsolete in the presence of tomorrow's flying disks we can only now call UFOs. Today's toxic nuclear and chemical environment will transform to more sustainable and healthy approaches to supplying food, energy, resources and medicine. The magic of today's science will become the practicality of tomorrow's technology. Research and development form the thin edge of a wedge of what's to come. Today's blueprints could become tomorrow's multitrillion dollar projects.

Consciousness as the Common Denominator of a New Science

Just as a television set would be magic to an aboriginal, the effects of consciousness that scientists are now observing are indistinguishable from magic. Resist this as we might, we are planting seeds for a new century that will not resemble the last. For example, imagine constructing electronic devices that respond to constructive human will. Imagine using the mind to produce energy, purify water, heal ourselves and others, travel and communicate across the universe, materialize and dematerialize things, and predict or modify the future. Imagine creating our own realities like the holodeck of Star Trek. Imagine restoring our inner and outer environments with the power of our minds—an elegant way to bring beauty and vitality back to the Earth.

This is the magic of consciousness. Once we begin to assemble the data, it should become easy for us to understand the power of our own consciousness, our ability to heal, the existence of an eternal soul, and prospects of communicating with intelligences other than human. In the last chapter, we saw how healing, survival of death studies and UFO/alien research are role models for consciousness science. In future sections, we will be looking at how the science of consiousness could provide a foundation for a new awareness of our greater being and as co-creators with Gaia of a positive future.

Experiments show that human intention can mysteriously alter the material world—demanding an entirely new view of physics. The research comes from competent scientists representing a wide range of disciplines. In this time of quickening, new fields are forming and growing so rapidly the present will soon seem to be a distant past.

My dualistic definition of consciousness involves a two-step process. The first is our intention to create something new in the universe. The second is how much the universe aligns with that intention. The degree there is alignment is the degree to which the miracles of consciousness can operate. Consciousness represents the most general case of the cosmic dance of cause and effect. It can either transcend or stimulate the four forces known to the old physics—electromagnetics, gravity, weak nuclear and strong nuclear. It can reach into dimensions beyond time and space. Consciousness can unite the universe or pieces of it instantaneously and even across time. Its medium of action appears to come from an invisible potential field embedded in the fabric of time and space.

In the last chapter we looked at immortality research and communicating with nonhuman intelligence as aspects of a new science of consciousness. The next five sections describe different types of experiments in zero point field research, quantum physics, parapsychology, materials science and cosmology. When combined with healing and the big picture of our greater nature, these could lead to a whole new future for humanity. The many repeated successes coming from these disparate approaches assure us that the new science of consciousness is not only promising; it is inevitable.

Explaining these concepts unavoidably involves principles of physics which might be foreign to some of you. If you have difficulty understanding the material, I recommend that you skim over it while pondering the essence of consciousness in action I have already described. Then you can pick up on the social implications described later in the chapter.

Experiments on the Zero-Point Field: Extracting Energy from the Vacuum

In Chapter 2 and Appendix I, I report experiments showing that energy can come from empty space. The condition required is that we accelerate an electromagnetic charge in the presence of the "zero-point field" (ZPF), called such because it still exists at a temperature of absolute zero, when all molecular motion ceases. We might also describe the ZPF as physicist David Bohm's "implicate order", which is the unmanifest state of matter and energy before it becomes visible. The ZPF might be an updated version of the nineteenth century physicists' dynamic ether— that invisible, ineffable medium through which matter and energy can be created or destroyed.

As examples of accelerating charges in the ZPF, physicists Bruce DePalma, Shiuji Inomata, Paramahamsa Tewari, Troy Reed and other investigators have measured excess output energies of rapidly rotating disks containing permanent magnets.[3] These scientists found that the power from these machines goes up as we increase the rotational speed, diameter and magnetic field strength. In principle, we might be able to rev up such a motor until over-unity power is obtained. At that time, we can unplug the machine so that it becomes a free-running generator of electricity. We will have tapped the zero-point field in much the same way as a paddle wheel dipped into running water produces usable energy.

The excess energy comes evidently from a stream of electrons jumping out of the void. When we think of the vacuum of space as a plenum of potential energy and charge, we see that energy is conserved in the larger sense. This answers the skeptics' accusations that "there is no such thing as perpetual motion."

The explanation offered by physicists Bernhard Haisch, Alfonso Rueda and Harold Puthoff, as reported in the peer-reviewed literature, has caused quite a stir in the scientific community. "The physical universe," they wrote, "is made up of massless electric charges immersed in a vast, energetic all-pervasive electromagnetic field. It is the interaction of those charges with the electromagnetic field that creates the appearance of mass."[5] In other words, matter may not be what we think it is.

They went on to describe the nature of this ZPF, which we have also seen to be the source of free energy. In his earlier presentations, Puthoff had frequently referred to the existence of this field from both experimental and theoretical perspectives. "The amount of energy making up the ZPF is enormous", the three authors said. "That energy, in the conventional view, is simply forced into existence by the laws of quantum mechanics... The existence of a real ZPF is as fundamental as the existence of the universe itself. "

" The idea that space could be filled with a vast sea of energy does seem to contradict everyday experience. The answer to the question lies in the utter uniformity (the same everywhere) and isotropy (same in all directions) of the field. There is no way to sense something that is absolutely the same everywhere, outside and inside everything. To put the matter in everyday terms, if you lie perfectly still in a tub of water at body temperature, you cannot feel the heat of the water." [5] What can you measure or detect from what?

I now quote from *Miracle in the Void* (Ref. 3, p.179): "The field becomes detectable only when a charge is accelerated through space. When that happens, the charge experiences an electromagnetic force as a resistance to the acceleration. Instead of the Newtonian concept of mass as a fundamental property of nature, it dissolves into being a force exerted on an accelerating charge. Collectively, these forces comprise energetic fluctuations that can be detected experimentally, for example through the widely reported Casimir effect. The forces are like a superposition over the seemingly featureless void of the ZPF."

"What does this mean? First, we can tap the zero-point energy by accelerating charges—for example, with a rotating magnetic disk or a specially conditioned magnet or crystal, or perhaps a cold fusion device or hydrogen gas cell. Secondly, since mass is really only an electromagnetic force, then gravity itself must also be an electromagnetic force acting on it; otherwise things wouldn't fall. Therefore anti-gravity propulsion becomes possible. 'To reinterpret Einstein's equation $E=mc^2$ ', Haisch et al write, 'we would say that mass is not equivalent to energy. Mass is energy.' "

This also implies that we can begin to truly unify the forces of nature without having to resort to such mathematically cumbersome formulations as string theory and gauge theory. I wrote: "Does this mean that consciousness itself is also an electromagnetic force, in order for it to be able to interact with the ZPF? I believe that may be so." (Ref.3, p.180).

If we pursue free energy from an old paradigm point of view and we keep living in isolation and suppression, we may continue to get sporadic results in new energy experiments. If we adopt a hypothesis that free energy is consciousness, the devices may then become a tool to fine-tune and augment the intentions of that consciousness. I know of several cases of free energy researchers who are apparently influencing their results through their intention. [3]

These heretical yet well-founded ideas coming from physicists provide the most elegant theoretical framework for understanding free energy, gravity, and perhaps consciousness itself. To further understand this relationship between the ZPF and consciousness, we next look in depth at experiments and phenomena within the fields of quantum physics, parapsychology, materials science and cosmology. I also explore quantitative aspects of the ZPF-consciousness relationship in Appendix IV.

Quantum Physics Inevitably Leads to Consciousness

During the early 1900s a number of physicists became baffled when they tried to subdivide the atom into more elementary particles. First they found that the energy states of one class of particle, the electrons, were discrete rather than a smooth continuum, and that the changes in energy from one level to another were sudden. Then they discovered that it was impossible to determine where a particle was going if they knew where it was; and just as impossible to determine where it was if they knew where it was going (named the uncertainty principle after Werner Heisenberg). Subatomic particles seemed to "flicker into and out of existence", as Haisch, Puthoff and Rueda recently described it. [6]

The physicists' confusion grew even greater when they found that the characteristics of the "particle" itself seemed to change simply by the act of observation. And more recent experiments showed that the particle is only a particle when it is observed; otherwise it exists only as a possibility, expressed as a mathematical probability in the form of a wave.

The most sensible, and perhaps only, explanation for this unexpected behavior is the observer effect, or our consciousness. In other words, we can change the physical properties of the observed, as in a mysterious dance of interaction. Suddenly gone is the Cartesian/Newtonian model of objective materialism, where the universe behaves like a gigantic mechanical clock whose properties can be understood by a dispassionate experimenter separated from the rest of nature. Gone is the notion that this clock is deterministically unwinding from an initial causal event about fifteen billion years ago called the Big Bang as the entire basis of our contemporary and future physical reality. Gone is the idea that all matter can be reduced to tiny inert billiard ball particles that zing around in time and space (sometimes called reductionism). And gone might be the "law" that the universe is gradually decaying in energy because of the thermodynamic requirement of entropy.

Instead we see a new kind of unifying and creative force between the observer and the observed. This is sometimes called idealistic monism, the perennial philosophy, the primordial tradition, the oneness, the alpha-and-omega, wholeness science, post-quantum theory, consciousness, mind, and, for some of us, God.

University of Oregon physics professor Amit Goswami has recently described the essence of the quantum mechanics which seems to fly in the face of objective materialism. [7] The experimental evidence points to three features endemic to idealistic monism: nonlocality, quantum leaps, and downward causation.

Nonlocality means that particles once together can still track each other's behavior out to great distances. Quantum leaps constitute the ability for particles to transfer information and energy at virtually infinite speeds—faster than the speed of light. Downward causation suggests that "consciousness is the ground of all being", in Goswami's words. Consciousness becomes much more than an anomalous departure from a basic belief in pure objective materialism, reductionism and determinism. "The fundamental tenets of material realism simply do not hold up", said Goswami. "In place of causal determinism, locality, strong objectivity, and epiphenomenalism, quantum mechanics offers probability and uncertainty, wave-particle complementarity, nonlocality and mixing of subjects and objects." Goswami concludes that all this adds up to the influence of consciousness.

If science itself could accept consciousness as the most general case of reality instead of trying to understand the universe in terms of its component parts in a clockwork universe, then some big changes are ahead. Reductionism and materialism then become but limiting cases of reality for those times when consciousness does not apply to any significant degree. In spite of the conservative biases of many physicists, the overall outlook is becoming more favorable towards basing reality on consciousness. "At a recent (American) Physical Society meeting", wrote Goswami, "one physicist was overheard to say to another: 'Anybody who is not bothered by Bell's theorem (a well-established principle of physics, based on expmeriments showing nonlocality) has to have rocks in his head.' Even more heartening, a poll of physicists at a conference revealed that a full 39 percent of the physicists assembled were indeed bothered by Bell's theorem. Since such a high percentage of physicists are bothered, we might well expect the idealist paradigm of physics to get a fair hearing."

Upward causation means that the essence of the physical universe can be understood by studying the most fundamental particles that make up atoms which make up molecules which make up cells which make up bodies, etc. Most physicists believe that each smaller part represents a more basic level of causation. This view still dominates the Western scientific paradigm and can be convenient at times, but seems to have reached a plateau of usefulness. On the other hand, causation spreading downward from consciousness has dominated Eastern thought. The new science which seems to be emerging is a science of consciousness, more in line with the major spiritual practices and religious beliefs of the world.

"Material realism cannot be saved", asserted Goswami. "...With idealist science , we have arrived at a science that has no entrance requirement, that excludes neither the subjective nor the objective, neither spirit nor matter, and thus is able to integrate the deep dichotomies of our thought."

Ironically, it has been the drive to find the fundamental and seemingly immutable particle that has led the Western approach to embrace the East. Rather than the whole as being a mechnical assemblage of parts with the dichotomy of object-subject, the whole *are* the parts. Particles are no longer the irreducible building blocks of nature, they are mysteriously flashing in and out of our reality, dancing with the observer, constituting a creative whole we call consciousness.

Many physicists are not aware of the importance of these results and instead turn toward the convenient technologies that emanate from adopting quantum theory. But nagging discrepancies remain: the wave/particle duality, the observer effect, nonlocality and spontaneous quantum leaps. The quantum paradoxes begin to disappear when we consider consciousness as the primary cause of the material world. Indeed, Goswami and some others of us trained in physics believe that consciousness literally creates the material world.

Our daily existence may be just another dream, the collapse of a wave of possibility into what looks and feels like concrete reality "out there". That reality seems to keep repeating because of our expectations of its manifestation. The discrepancies remain. These habitual patterns appear to come from a deep-seated but erroneous belief in the primacy of an objective, material universe which we cannot change except by the brute forces of gravity, traditional electromagnetics, nuclear repulsion and atomic decay.

During his later years, Einstein attempted in vain to come up with a unified field theory which would combine the four forces. He was also troubled with the paradoxes of quantum mechanics, but could never arrive at a reconciliation. If physics were to honestly address all valid observations, it would have to become more all-inclusive. In other words, the currently accepted theories of physics fall far short of explaining the growing experimental anomalies.

Goswami is not the only theoretical physicist who is convinced consciousness must be brought into a faltering physics. Cyberphysicist Dr. Jack Sarfatti, a unique contemporary of mine, blazes new theories in large daily morsels on the internet. [8] He is proposing a post-quantum theory, based on his attempt to merge the apparent contradictions of current quantum models. He also draws on the results of Dr. Thomas Phipps, who has modified relativity theory to fit in with quantum mechanics. Sarfatti presents a model of consciousness in which not only does mind act on matter but matter reacts back onto mind. He uses this to explain the observer effect in quantum physics and psychic effects as well. In a more formal way, he underscores my definition of consciousness, which is the result of the degree of alignment between our intention and the intention of that being acted upon.

There are many more physical theories of consciousness, too numerous to mention here. It is unfortunate that not only is currently accepted physics riddled with contradictions, but both old and new concepts themselves are often difficult for the lay person to understand, because the terminology, mathematics and models are complex. Even with all my physics background, I need to take out a lot of time to begin to grasp the subtleties of the theories. It is easy for some of us to want to give up, to walk away from all this intellectual rigor.

Yet physics is still a very powerful tool, no doubt about it. Existing theories do explain a variety of phenomena and predict future events. They give us confidence to understand how nature behaves, how we can interact with it, and how we can create everything ranging from microelectronics to lasers to computers to communications to aircraft and space vehicles. But the emerging theories which seek to explain the anomalies of consciousness, added to accepted results, will no doubt spell out a future that would make even the postulates of an Einstein, Planck, Heisenberg, Bohr, Fermi, Dirac, Bohm or Schrodinger appear mundane. Such is the cutting edge nature of science.

Many other scientists outside of physics are also uncovering additional evidence for the influence of the observer on the material world, beyond the ZPF and quantum experiments. A third area which is growing rapidly in credibility consist of observations of parapsychologists and other academic researchers of the paranormal. These anomalous happenings violate the laws of physics as we know them. They open ever wider a Pandora's box leading inevitably toward the primacy of consciousness in understanding the physical universe.

Parapsychology Experiments and Experiences Also Prove Consciousness

While quantum mechanics and zero-point energy extraction reveal a bizarre behavior of "matter" at the microscopic level, the observer effect also applies to the macroscopic (visible) world. These results come from a series of alarming investigations into mind over matter (psychokinesis), remote viewing, precognition, near-death experience and other phenomena clearly foreign to the materialist yet very real. Human operators interacting with machines through their intention can often create, destroy or move electrons in repeatable trials, just like the observations of the quantum physicists and zero-point field researchers.

I described in my book *The Second Coming of Science* the important experiments carried out by Princeton University scientists Robert Jahn and Brenda Dunne. [29] They have devised an experimental protocol which can measure the psychokinetic effects of human operators on random event generators (REGs) of binary numbers. These are black boxes which, in the absence of any operator intention, produce a random stream of zeros and ones at the rate of two hundred numbers per second. By integrating results over several hours, they have found for most operators a statistically significant and moderately repeatable "psychic" effect both in and opposite the direction of intention, with a preponderance of results towards intention. Typically, operators have produced non-random effects of a few parts per thousand. One doesn't need to be a gifted psychic to produce results. Statistically we are all psychics.

Once I may have created in my own mind erroneous positive results in an experiment I had performed. [2] As an enthusiastic astronomy graduate student at the University of California at Berkeley, I had sought an optical effect coming from the planet Venus which could reveal the presence of hexagonal water ice crystals in its cloud tops. Indeed I found something. But three years later, when I was an assistant professor of astronomy at Cornell University, a graduate student and I didn't find it. As a cautious scientist, I fortunately hadn't come to any positive conclusions about the presence of ice the first time and awaited verification. From both sets of observations we set an upper limit to the content of hexagonal ice crystals in the Venus cloudtops. Soon, another scientist discovered sulfuric acid droplets there instead. Caution in science is wise in cases like these; I object strongly only when scientists pretend to know everything and debunk new fields unfamiliar to them.

What happened to the ice in the Venus clouds? I believe I influenced the electronics of my photometer at the telescope to respond to my intention. These anomalies came up in many other experiments, as we shall see. All this forces the question, how many wish-fulfilled results come from our own consciousness rather than physical reality "out there"? Lacking the objectivity demanded by one scientist can become grist for the mill for another.

Subsequent experiments at the Princeton Engineering Anomalies Laboratory have shown ways of amplifying psychokinetic action. Brenda Dunne has reported some preliminary results indicating psychokinetic effects approximately six times greater for bonded couples. More recently, Roger Nelson, also at Princeton, and Dean Radin at the University of Nevada Las Vegas, have confirmed that even greater order can be produced from focusing group energy in the presence of an REG. [10,11]

These new results come from two teams of experienced experimenters in a cautious academic settings with peer review. By placing "field" REGs in various group settings, both experimental teams found repeatable patterns of extrasensory group coherence, particularly during moments of strong spirit, or bonding. The coherence was greatest when humor, ritual, inspiration, or large television audiences were involved. On the other hand, a routine scientific board (bored?) meeting yielded continuing randomness. Interestingly, the groups did not have to focus on the REG itself; the object of focus could be something else. The prospect of using field REGs as group monitoring feedback devices suggests that we can find the best conditions to "pump up" group energy into a highly coherent state, perhaps even into a "free energy" domain. If you would like to learn more about possible quantitative relationships between amplifying consciousness and zero-point energy extraction, I refer you to a short technical treatise in Appendix IV.

Some of us have difficulty understanding these complex scientific results coming from black boxes, and so would prefer seeing or experiencing dramatic demonstrations of psychic action. Both approaches verify the reality of consciousness. Besides observing gifted psychics, I have also taken up spoon bending. Through various aikido and breathing exercises, I have taught this during the culmination of workshops at the Esalen Institute in California, the Findhorn Foundation in Scotland, and elsewhere. In a two-to-three hour session, about eighty to ninety per cent of my students are able to bend a spoon, where they are clearly convinced it was primarily done with the power of their minds, rather than physical force. Interestingly, the middle-aged intellectual men performed most poorly, a credential role reversal. [2] Firewalking represents another valid and personally more challenging confrontation with consciousness.

Diehard materialists nevertheless question all this. Dean Radin's recent book *The Conscious Universe* should lay to rest those doubts once and for all. [11] I highly recommend this brilliant and understandable book. In it, he synthesizes thousands of disparate experiments in parapsychology which not only prove the reality of consciousness existing beyond time and space, but also lead to certain attributes that will impact any emerging theory of consciousness.

Radin's confidence can be bolstered by the statistics of large numbers in hundreds to thousands of experiments, as reflected in a "meta-analysis". "In a given experiment", he writes, "the raw data points are typically the participants' individual responses. In meta-analysis, the raw data points are the results of separate experiments...Today meta-analyses have exploded in popularity because the behavioral, social and medical sciences were all in the same boat: they needed a method of formally determining whether the highly variable effects measured in their experiments were replicable." (p.53)

For example, Radin's look at thousands of studies on telepathy and perception at a distance show significant "hit rates". The studies also reveal that about one per cent of the population have an innate psychic talent, and no amount of training seems to help the others much. Regardless of psychic ability, the analyses indicate that there is greater performance when operators remotely view objects and events by free choice rather than by constrained choices, such as concealed cards. Hypnosis can further enhance the effects. These findings are consistent with my own experience and with experiments I did with polygraph scientist Cleve Backster. [1] In one set of trials, the electrical activity of my at-a-distance white blood cells time-correlated well with my experienced and recorded emotions.

Even more mind-boggling are Radin's results of perception over time. Several independent experiments clearly show the abilities of many people to predict and perhaps even influence events in the future. Through measuring their electrodermal activity, Dr. Dick Bierman found that many subjects were able to anticipate an emotionally charged image seconds before it was actually presented. [11]

An extreme example of precognition comes from Ted Owens, the late American psychic. He was able to predict—and sometimes even claimed to produce—freak thunderstorms, lightning strikes, hurricanes and sporting event results. [12] While Owens was for the most part considerate of others, there were times he followed through on reprisals against those whom he didn't like. This leads to a serious issue. Our potential to play God serves to remind us of the double-edged sword regarding ethics in the twenty-first century. To re-inherit the Earth, we are going to need to follow the authentic Golden Rule in all scientific inquiry: Do unto others as you would have them do to you.

These very significant results coming from parapsychology must be included in a theory of consciousness, whose actions can spread over both space and time following natural laws that far transcend what we now know in science. "At the turn of the twentieth century", Radin writes, "imaginative scientists were slowly becoming aware of radical new theories on the horizon about space, time, matter, and energy. Some sensed, correctly, that developments such as relativity and quantum theory would radically alter our understanding of reality itself. Almost a century later, the impact of these discoveries is still reverberating throughout science, technology, and society. As the twenty-first century dawns, astounding new visions of reality are stirring." (Ref. 11, p.303)

The Influence of Human Intention on Water and Living Matter: One Man's Amazing Synthesis

One of the brightest and most articulate experimenter-theorists in consciousness is professor emeritus William Tiller of Stanford University. This gentle man whom I've recently had the pleasure of meeting, has provided convincing data on the influence of human intention on the physical and chemical properties of water and living matter. Even more remarkable, Tiller and his colleagues were able to delay the transmission of the intention through the mediation of a simple electronic device. [13] They used experienced meditators to implant their conscious signal into the device, which recorded their intention.   They later transmitted that signal into water, for example, to alter its degree of acidic or alkaline content.   Some repeated trials yielded 1.0 pH unit, or ten times change in the acidity of the water coming from the relayed intention of the meditators.

Tiller has presented a model for consciousness I find to be both brilliant and useful. [14] I often prefer models to formal theories, especially at this early stage in the development of consciousness science.   While both have powerful predictive qualities, physical theories usually build on existing theories, and often demand a rigorous understanding of mathematical physics.   Models, on the other hand, are attempts to build a generalization that is consistent with all the relevant data.   Models are the simulation of a universe which shares all its observed properties, not your pet theory.   True, the theoretical physicists can have important insights leading to later developments, but the systems engineers can often build models beyond the purview of the physicist before the physics becomes more developed.   Systems engineers got us to the Moon, based primarily on physical theories advanced by Isaac Newton 300 years ago.

Systems engineers are often the first people to point to trends for the future.   This kind of thinking came from the Club of Rome's prophetic 1968 book *The Limits to Growth*.   This study looked at five key parameters in modeling humanity's future with respect to the resources of the Earth: food, raw materials, energy, pollution and population.   They correctly predicted an environmental crunch which is obviously upon us now.   This important work also helped to create the environmental movement of the 1970s.   Systems engineering will undoubtedly help us design a positive and sustainable future.

In a sense, the new Apollo program has already begun.   Thomas Bearden and Moray King are systems engineers who have provided fresh insight into the characteristics of zero-point energy. [2,3] The visionaries Werner von Braun, Sir Arthur C. Clarke, Gerard O'Neill, Buckminster Fuller, Barbara Hubbard, and Willis Harman have published works in the finest tradition of systems engineering.   Their ideas deserve a new look. Tiller's own background is an eclectic blend of engineering, materials science and holistic medicine, so he is a natural candidate to address the broad question of consciousness systems modelling.

In his model Tiller begins with the familiar physical realm which includes the three dimensions of space and one dimension of time. But in order to explain the actions of the mind on matter he proposes a second, unseen domain that has certain general properties that can be deduced from the data of many classes of interactions: quantum experiments, parapsychology, zero-point energy research, holography, chemistry, biology and healing science, to mention a few. [14] The two other realms he suggests are more spiritual in nature.

Tiller calls our familiar space-time domain direct or D-space and the unseen inverse space-time (frequency) domain as reciprocal, or R-space. The mirrored reciprocal space doesn't only involve inverse space and frequency. Some other features of R-space are negative mass, energy, entropy and temperature; velocities greater than light; levitation as opposed to gravitation; homeopathic versus allopathic medicine; and right versus left brain.

The model uses the familiar Fourier transform of mathematical information and action back and forth between the two domains. Space-time becomes converted to inverse space-frequency and vice versa. This also follows the holographic model of the universe first proposed by the late physicist David Bohm, and expanded by the late Michel Talbot [15] and astronaut Edgar Mitchell, founder of the Institute of Noetic Sciences. [16] This general approach can also explain how many of the phenomena that seem to act outside of space and time can impinge on the moment. The unseen realm is resonating with the seen to create the observed reality in this moment at this place.

Tiller's model is also consistent with one proposed by the late Japanese engineer Dr. Shiuji Inomata, president of the Japan Psychotronics Institute and for thirty-five years an employee of the Japanese Government at the Electrotechnical Laboratories near the Tsukuba space city facility. He was host of two Tokyo conferences on Consciousness, New Energy and New Medicine in 1996 and 1998. Dr. Inomata has presented a triad between mass, energy and consciousness as a three-way interaction. [3] He suggested that existing science has concerned itself only in the relationship between mass and energy. But if we add consciousness as a third factor, we find physical equivalences between consciousness and both mass and energy analogous to the famous Einstein mass-energy equation, $E=mc^2$.

Tiller explains a wide range of observed data with some understanding to the informed layperson. He also provides sometimes poetic descriptions of many demonstrations of paranormal phenomena.

Regarding materialization, he writes, "Imagine a pond surface on which little 'skimmerbugs' are moving about on the surface. They have a small body and many long legs, and resemble Abbot's Flatlanders in that they perceive only two dimensions. Suppose you have a skimmerbug skimming across the pond near you and you put your leg in the pond. To the skimmerbug, that is a materialization event. You then take your leg out of the pond and, to the skimmerbug, that is a dematerialization event. If you think of the number of ways in which your three-dimensional phenomenon can penetrate its two-dimensional perception frame, you can begin to appreciate how difficult it will ever be for the science of skimmerbugdom to be able to produce a proper scientific explanation of the phenomenon. We see here that one's scientific prowess is limited largely to the level of the society's operational perception." [14]

I highly recommend Tiller's model to serious students of consciousness science. "The break with the past will come", he writes, "in part, by accepting that, like light and sound, our present band of cognition gives us a window on only a very small portion of Nature's total modes of expression...Becoming aware of these larger aspects of general nature of ourselves is a major step towards full self-empowerment, where we have grown in consciousness enough to meaningfully influence the properties of matter around us. By then, we will know what true balance means and can set about restoring it in the world."

We give thanks to William Tiller's important contributions, which provide an excellent formulation for a new science of consciousness. Perhaps the most powerful results will next come from combining the new models of the engineers with the theories of open physicists such as Amit Goswami.

The Cosmology of Consciousness

Some of these amazing experiments in consciousness can lead us to re-examine the universe in profoundly new ways. For example, we have Einstein's theory of Special Relativity. One prediction of this theory which has been experimentally verified is that the passage of time converges to zero when one approaches the speed of light. This phenomenon of time dilation gives rise to the famous "clock paradox" in which hypothetical travelers on a round trip from Earth into space at nearly the speed of light could return being younger than their own children. Decades may have passed on Earth while perhaps only months were clocked on board the spacecraft.

Based on this principle, author-scientist Peter Russell presented the following thought exercise in his recent book *A White Hole in Time.* [17] We all know, he said, that it takes nine minutes for us to receive light from the Sun because of the measured finite speed of light (c=300 million kilometers per second). But Russell poses an intriguing and inevitable consequence of relativity when we change our perspective to that of light itself.

"As far as light is concerned," he writes, "the moment it left the Sun is the same moment it arrived at my eye. From its perspective there is no time interval. This coincides exactly with my experience. The realm of consciousness and the realm of light would seem to share the same experience of now."

In other words, if you are light, you are everywhere at once. As a particle of light you can travel from the Sun and instantaneously move through an Einsteinian curved universe to all locations like a ball of string. Any meaning to the idea of both time and space disappears, because the sunlight floods every point in the universe at the same time (from its point of view). The same is true of all light in the universe. As light, we have entered a new domain of reality.

Could the collective light of the sun and other energetic sources in our universe also be at every point in space at all times? Is this possibly related to the zero-point field? Could it be, the field itself is that light which is everywhere at once? Are we inhabiting two realms simultaneously—the familiar space-time domain, and the less familiar one of consciousness and light? Or shall we call it a new dimension? Tiller's model helps make all this magic more understandable.

My own hunch about the new directions of physics will include a good look at the heretofore elusive zero-point field and its interaction with consciousness, as I describe in Appendix IV. Also we will need to further develop post-quantum theory, which deals with reconciling the observer effect, mind-over-matter, and relativity. As the concepts begin to fuse, they could possibly become the breeding ground for new technologies that might solve our most pressing problems of the environmment.

Consciousness Reaches Across the Board in the Academic Disciplines

In recent years academic philosophers and theologians, prodded on by the amazing results of quantum physics, have also taken a new look at consciousness. In a groundbreaking book *The Conscious Mind*, David Chalmers, a philosophy professor at the University of California Santa Cruz, argued cogently that the mystery of subjective self-awareness cannot be expained in its entirety by objective materialism. Consciousness, he says, needs to be reinserted into the practice of science and philosophy. "It seems to me", he wrote, "that to ignore the problems of consciousness would be antiscientific... Materialism is a beautiful and compelling view of the world, but to account for consciousness, we have to go beyond the resources it provides."[18]

Chalmers opened his book with these key perceptions: "Consciousness is the biggest mystery. It may be the largest outstanding obstacle in our quest for a scientific understanding of the universe. The science of physics is not yet complete..." I would add to his remarks that the obstacle of consciousness perceived by the materialists can become an opportunity when we look at the impressive data coming in from the consciousness sciences I have mentioned in this chapter. Sometimes, mainstream scientists and philosophers can accept that only the well-established principles of quantum physics can lead to consciousness. This is because quantum physics has slowly gained general respect in academic circles, whereas parapsychology, the zero-point field, the Tiller results and other anomalous experiments are very young yet in the Western mind. But I am convinced that we shall see changes very soon, as the paradigm unfurls.

Chalmers is not alone in the conviction that consciousness deserves a closer look. Hundreds of scientists, theologians and philosophers gather each year in Tucson, Arizona, to discuss and debate consciousness. They have set up an interdisciplinary team to establish the Journal of Consciousness Studies: Controversies in Science and the Humanities. Contributors include the nobel laureate Francis Crick, astrophysicist Roger Penrose, philosopher Ivan Illich and the late Willis Harman, former president of the Institute of Noetic Sciences. Many of the articles and papers reflect a materialistic bias, but we can see progress coming from many fronts.

I recently spoke at the International Conferences on Science and

Consciousness in Albuquerque, New Mexico in April 1999 and April 2000. Other speakers included Peter Russell, Dean Radin, Jahn Hagelin, Brian Swimme, Edgar Mitchell, William Tiller, Stanley Krippner, Jeffrey Mishlove, Elisabet Sahtouris, Larry Dossey and many others. Over 500 people attended each conference.

We also have the philosophical field of metaphysics, with many pioneers appearing on the scene outside of mainstream academic circles. For more than a century, Alice Bailey, Manly Hall, Charles and Myrtle Filmore, Mary Baker Eddy, Ernest Holmes and many others have established what later became large organizations dedicated to the study and practice of consciousness and the healing power of prayer. Included are the Philosophical Research Society, the Theosophical Society, Unity Churches, Churches of Religious Science, the International Institute of Integral Human Sciences, the California Institute of Integral Studies, and the Sivananda yoga ashrams. Several academic philosophers are also involved in the new sciences, including Stanley McDaniel, Michael Zimmerman and Michael Grosso. Eastern philosophers, mystics, yogis and shamans have known about consciousness for a very long time. Perhaps the most significant contributions have come during the early twentieth century by the Indian philosopher Sri Aurobindo.

Since the time of Carl Jung, consciousness has also been the cornerstone of many innovators coming from the discipline of psychology: J.B. Rhine, Stanley Krippner, Stanley Grof, Ram Das, Russell Targ, William Roll, Bernard Grad, Robert Morris, Erlandur Hearaldsson, Lee Pulos, Charles Tart, Jeffrey Mishlove, Brenda Dunne, Ken Ring, Dick Bierman and Dean Radin, to mention just a few, whose work is reported widely here and elsewhere. Consciousness has become a cornerstone for psychiatrists John Mack, Brian Weiss, Elizabeth Targ and the late Timothy Leary.

Towards a New Science of Consciousness

A number of organizations of new scientists have emerged over the past two decades to focus on consciousness as their central theme: the Institute of Noetic Sciences, the International Association for New Science, the American and Japanese Psychotronics Institutes, the Society for Scientific Exploration, the Scientific and Medical Network in Europe, and many others. Several outstanding publications have come out of these organizations. Probably the most prominent is the peer-reviewed *Journal of Scientific Exploration*, which has published many of the results decribed in this chapter.

Several other experiments also point to the reality of consciousness. We have seen that healing with the mind, communicating with the dead and other beings, past-lifetime recall, near-death experience and UFO studies give clear indications that the energy fields of consciousness are very real and are not necessarily respecters of space and time. Consciousness is also an important factor in all living systems. For example, we have Cleve Backster's astounding data on telepathy and biocommunications [1]; Rupert Sheldrake's experiments on various species showing intraspecies communications that cannot be explained entirely by heredity and the environment [19]; and the miraculous healings many individuals have reported widely in the literature. All of life seems to possess energy fields of consciousness that transcend time and space.

The recent stunning developments in defining the human genome reveal a mysterious complexity in the molecular structure of all life. The human body-building genetic code comprises only about one per cent of the total. At a molecular level, there seems to be nothing special about being human: even the chromosones for pigs, plants and yeast are almost as complex as ours. This result is very humbling for our anthropocentric selves, and is consistent with Backster's experiments that plants and yeast and humans alike show electrical activity in response to human intention.

The DNA molecule itself might act as a transducer for consciousness [3, 20], perhaps explaining some of the unknown function of the genetic code. Also some of the DNA strands include signs of mutations within ancestors. Many portions of the genome have unknown origins, which could conceivably include historical mutations from genetic manipulations by extraterrestrial beings. We might be able to test the hypothesis that alien visitors have occasionally been cross-breeding with humans. Perhaps we may have descended from more than apes, a vexing thought for mainstream geneticists already puzzled with the genome results.

The paradigm shift towards an age of consciousness doesn't end with physics and biology. Chemistry is also in for an overhaul, with the revising of the ancient art of alchemy. In Chapter 2, we saw that cold-fusion technology can be used to transmute elements at room temperature. This will enable us to remediate radioactive and other toxic wastes. Nine American patents have already been granted to inventor James

Patterson to begin the task of detoxifying our chemical environment.

Earth science could also be in for a big change. We are beginning to understand that the Earth might behave as a superorganism, whose unified complexity, intelligence, and level of self-organizing cooperation rivals those of the human body itself. [1,21] This new paradigm of a living, conscious planet might help us decipher the enigma of global warming and climate change, diagnosed perhaps as a worldwide fever. The tools of consciousness could be at the forefront of our goal of moving towards a sustainable environment. "The idea of the world as an organism has been called the Gaia hypothesis", writes Radin, "named after the mythical Greek goddess of the earth. Do field-consciousness effects suggest that there may be a mind of Gaia?"

"....under exceptional circumstances—during worldwide, live television broadcasts, for instance—when many minds are focussed on the same object, unbeknownst to us a grand alignment occurs. During these brief, shining moments, the billions of individually glittering minds reassemble into a whole, and the unity of Gaia's mind becomes brilliantly manifest. At such uncommon times (but becoming more common every day), Gaia in effect awakens, and we see this reflected in our random systems because they suddenly start behaving in statistically unexpected ways." [11] Our coming together as newly enlightened beings could lead to planetary healings. After all, our connected consciousness make us one and the same.

The Coming Consciousness Revolution

I am convinced that consciousness science will allow us to solve our deepest environmental challenges in elegant new ways. For example, we might be able to clean up our waterways through our intention. Our work has barely begun, but a direction is there. We may soon be able to examine the symbiotic relationship between energy coming from the zero point field, consciousness and group intention. I invite you and interested colleagues to contribute your insights into these most important experiments as the vision of a new science of consciousness unfolds. One great advantage of proceeding this way is that we don't need to build multibillion-dollar particle accelerators, space probes, guided missiles, atomic warheads and powerplants, orbiting telescopes or fusion reactors to do consciousness science. Many of the experiments and concepts are simple enough for an informed layperson to address.

Acknowledging and studying the zero-point field, quantum and psychic interactions—which may be the source for understanding the genesis of free energy, gravity and consciousness—is an action akin to placing the sun in the center of the solar system. The epicycles (the analogue to materialism) begin to disappear. They vanished entirely when later refinements such as Kepler's and Newton's Laws were brought forward.

Likewise, the current discoveries inevitably lead to a revolution in the sciences which will spread quickly. Combining these fresh ideas along with some others now budding can bring us into the new scientific paradigm of consciousness and the ZPF (Appendix IV). Perhaps this *is* the Consciousness Revolution. So here is the next step to re-inherit the Earth: Create and support a new science of consciousness.

The miracles of today will become the commonplace science of tomorrow. Maybe we can all become empowered to extract energy and matter from the void. These activities could bring us into a higher dimension which only awaits our acknowledgement and exploration. We may then be able to learn to resurrect ourselves from a mortal and finite existence. Then we will become more fully conscious of our place in a universe made alive and connected with who we are, and in the process, learn how special all life on Earth is. This is what the new paradigm is about—transcending our self-imposed imprisonment by materialism, determinism and reductionism. We need to look at all this in light of the challenges of the twenty-first century.

I said in the conclusion of my last book, "the current discoveries inevitably lead to a revolution in the sciences which will spread quickly. Combining these new ideas along with some others now budding, is bringing us into the new paradigm, into refining our understanding of consciousness and the ZPF. Perhaps this *is* the Consciousness Revolution...."

"The miracle in the void is that we can all empower ourselves to create beautiful new worlds, magnificent new universes. When we begin to resonate with the majestic and ubiquitous reservoir of pre-energy and pre-matter in the zero-point field, we will all becomes healers, clairvoyants and magicians. We can at lest have peace, harmony, love and joy. Science is telling us that clearly, based on irrefutable experimental, theoretical and personal evidence. I invite you to trust the process and to walk with me through the visible into the invisible." [3]

Perhaps we can now envision a sustainable future in which a

blend of consciousness and common sense is creating powerful and benign new technologies whose time has come. But we are going to have to support the science of consciousness, hopefully through public participation. This field is now in its infancy, but may we hope it will grow rapidly once the research moves ahead. We don't only need a Los Alamos for new energy, we shall need to create a new Apollo program for consciousness science.

In the last chapter we looked at the successes of consciousness medicine, which sets an example as the first multibillion-dollar industry using the tools of consciousness. Great strides have been made in remote healing that are not only paying off; they provide the perfect role model for healing the Earth. It would not be out of the question that we could end pollution and ensure a sustainable future simply through the focussed intentions of groups of people.

We are in a global spiritual crisis which demands that we remove our veils of denial and enter a new science of consciousness, exploring our potential to heal, our eternal nature, and our membership in a cosmic community of sentient beings.

References for Chapter 6

1. Brian O'Leary, *Exploring Inner and Outer Space*, North Atlantic Books, Berkeley, 1989.

2. Brian O'Leary, *The Second Coming of Science,* North Atlantic Books, Berkeley, 1993.

3. Brian O'Leary, *Miracle in the Void*, Kamapua'a Press, Kihei, HI, 1996.

4. Beverly Rubik, *Life at the Edge of Science*, The Institute for Frontier Science, Philadelphia, 1996.

5. Bernhard Haisch, Alfonso Rueda and Harold E. Puthoff, "Beyond E=Mc²", *The Sciences*, (November/December, 1994) and *Physical Review A* (February 1994).

6. Bernhard Haisch, Alfonso Rueda and Harold E. Puthoff, "Advances in the Proposed Electromagnetic Zero-Point Field Theory of Inertia", presentation to the 34th AIAA/ASME/SAE/ASEE Joint Propulsion Conference and Exhibit July 13-15, 1998, Cleveland, Ohio.

7. Amit Goswami, *The Self-Aware Universe: How Consciousness Creates the Material World*, Tarcher/Putnam, 1995.

8. Jack Sarfatti, cyberphysicist, www.stardrive.com.

9. Robert Jahn and Brenda Dunne, *Margins of Reality*, Harcourt Brace Jovanovich, San Diego, CA, 1987.

10. Roger Nelson, G. Bradish, Y. Dobryns, B. Dunne and R.Jahn, "Field REG Anomalies in Group Situations", *Journal of Scientific Exploration,* volume 10, pp.111-42, 1996.

11. Dean Radin, *The Conscious Universe*, Harper San Francisco, 1997.

12. Jeffrey Mishlove, *The PK Man*, Hampton Roads, Virginia 2000.

13. William A. Tiller, Walter E. Dibble, Jr., and Michael J. Kohane, "Exploring Robust Interactions between Human Intention and Inanimate/Animate Systems", presented at "Toward a Science of Consciousness—Fundamental Approaches", May 25-28, 1999, United Nations University, Tokyo, Japan.

14. William A. Tiller, et al *Science and Human Transformation*, Pavior Publication, Walnut Creek, CA, 1997; *Conscious Acts of Creation: The Emergence of a New Physics*, ibid, 2001.

15. Michael Talbot, *The Holographic Universe*, HarperCollins, New York, 1991.

16. Edgar Mitchell, *The Way of the Explorer*, Institute of Noetic Sciences, 1998.

17. Peter Russell, *A White Hole in Time*, Harper San Francisco, 1992.

18. David J. Chalmers, *The Conscious Mind*, Oxford University Press, Oxford and New York, 1992.

19. Rupert Sheldrake, *Seven Expreriments that Could Change the World*, Riverhead Books, New York, 1995.

20. David A. Ash, Science of the Vortex, *The Light University*, 4 Western House, Station Road, Totnes, Devon TQ9 5LF, United Kingdon, 1993.

21. James E. Lovelock, *Gaia, A New Look at Life on Earth*, Oxford University Press, 1979.

7

Adopt a Manifesto for Sustainability:
A Declaration of Interdependence

"Our current generation is committing Treason against future generations by destroying our global environment."
-Norman Cousins

"Science has made unrestricted national sovereignty incompatible with human survival. The only possibilities are now world government or death."
-Bertrand Russell

I CANNOT HELP but feel pretentious about drafting a binding document for the whole world, asserting what must be done. Adding my own small contribution to what could be eventually put into the collective, whether it be in the form of a cultural creative's bill of rights, manifesto, resolution, articles, declaration, draft constitution or whatever, can feel like reinventing the wheel. Many such documents already exist. There are probably many I do not even know about. In this short chapter, I try to take the best of some of these ideas to come up with a succinct statement about our common purpose, in the hope of stimulating thought and action.

Later drafts of this document will further incorporate the spirit of what has been expressed many times, in many ways. Among the philosophies that make most sense are those of Ervin Laszlo of the Club of Budapest, and Benjamin Ferencz and Ken Keyes in their book *PlanetHood*. Laszlo recently drafted a *Manifesto on the Spirit of Planetary Consciousness*.

It captures many basic ideas behind the formation of a world democracy. After acknowledging the breakdown of the human condition and natural environment globally, Laszlo expresses a call to responsibility for creating a global community based on the spirit of sustainability. "Planetary consciousness," he writes, "is the knowing as well as the feeling of the vital interdependence and essential oneness of humankind, and conscious adoption of the ethics and ethos that this entails." [1]

Elsewhere, Laszlo says: "Evolving the human spirit and consciousness is the first vital cause shared by the whole of the human family. Responding to the challenge of...sustainability calls for another Apollo mission on the plane of culture. Creative people in all relevant spheres of culture need to be brought together and encouraged to put their insight to work in the joint human interest."[2]

When we combine these basic principles with those in *PlanetHood*, that we all have the right to "live with dignity in a healthy environment free from the threat of war"[3], I believe we have the basis for a working document to be considered as a manifesto which could be signed by up to 120 million cultural creatives worldwide, with more joining us later.

The larger challenge is to create a constitution for a global green republic whose powers to make laws, enforce them and adjudicate disputes in issues of sustainability, peace and human rights, would exceed those of existing nation states and corporations. At the same time, the new government will need to give broad discretion to local governments to carry out its goals. This process could be long, arduous and contentious, as the forming of the American democracy demonstrated.[3] The framers of the U.S. Constitution deliberated for over a decade after they declared independence. It was not easy for the then-sovereign states to give up their special interests, and so there were bitter controversies which took decades to resolve. Yet prior attempts to form the union as a confederation of states had failed. It was only when the founding fathers wrote a binding constitution that a union could be formed, united in purpose.

We have the same opportunity now to save our planet from almost certain destruction. The American experience is apt: now we will have to form a new union based not only on the principle of independence, or personal sovereignty, but on the principle of interdependence globally. We want to recreate sovereignty among individuals and communities and to create anew a global sovereignty, a cosmic sovereignty.

When the American states agreed to give up their military power and to surrender many other sovereign powers, a national democratic government, with checks and balances, could begin to represent a federation which effectively stood up to the tyranny of the English monarchy. We shall need to do a similar thing with nation-states and corporations: they too will have to give up much of their sovereignty. They can choose to join the new team too, which would empower us all the more. There are no losers in this except for vested interests, mostly financial. We have seen in this book that the tyranny of nation states and large corporations must be overcome, but so many positive things can happen in a global green republic that its popularity will overwhelm the old paradigm.

The transition will be messy, as always, but our fear of it cannot stop us. I am opposed to violent revolutions, and I believe those will be unnecessary. While the new democracy grows, so nation-states and large corporations will shrink in power. As a start, we will need to offer an olive branch to our adversaries as individuals in the now movement. If they don't want to participate, we could apply the new Golden Rule in the form of golden parachutes that would ease them down from power.

Ferencz and Keyes believe that we can now form the new government. Their insightful and inspiring book include quotes from disparate individuals, all of whom believe that some sort of world government will be necessary for our survival: Robert Muller, Norman Cousins, Mikhail Gorbachev, Elliot Richardson, Emery Reve, Peter Ustinov, Albert Einstein, John Kennedy, Theodore Roosevelt, Woodrow Wilson, Winston Churchill, and, yes, George Bush, Senior. The proposal of *PlanetHood* is to empower the United Nations and/or to create an international constitutional convention for a new global democracy whose jurisdiction would include provision for a sustainable, just and peaceful future. In Chapter 4, I expressed my own preference for starting something entirely new and to use the resources of the U.N. as a means, not an end to world government. The situation calls for leaving behind all vested interests and acting as equals.

I envision a global constitutional convention to enact the governmental body to follow the principles of a manifesto similar to that presented below. As in the American case, the convention would thrash out the major issues which will no doubt threaten some aspects of corporate and national sovereignties. So be it. For we should never lose sight of the reason why we were brought together in the first place. In the end, it will be the sovereignty of individuals and our environment which must be upheld, not the special privileges now granted to economic government.

A Manifesto for Sustainability: Declaration of Interdependence

Whereas the ravages of humanity upon the natural environment and upon one another are no longer acceptable for our future and that of the planet and other living beings,

Therefore be it resolved that we the citizens of Earth declare our right to live with dignity in a healthy environment free from the threat of war and from the suppression of new knowledge.

To these ends, we now form a global republic whose power shall guarantee the following results: 1) a sustainable environment; 2) the end of war; 3) the right of every person to food, shelter, meaningful work, health care, justice and free access to knowledge; and 4) an ethic which ensures that these principles shall endure.

Is That All?

In essence, the foregoing draft embraces what we need to do. It will be up to enlightened constitutional scholars and others to come forward and to make their contributions. The titles of the chapters of this book, as listed in the table of contents, state some of the basic actions we will need to take under the new government. I make no pretense that this process will be easy. As in the case of the formation of the American democracy, many long hours, days and years will be spent debating, deliberating and ultimately transcending the resistance of sovereign groups whose power will be inevitably subsumed under the new democracy. We can realistically expect that the vested interests of economic government will compel them to fight us all the way. For example, we can imagine that those in power will be placing obstacles in the way if a global law were passed mandating an 80% reduction in greenhouse gases worldwide by 2015. But, unified in purpose, we shall succeed in the end; that is part of the design. The only way we can fail is not to do this at all, and that is tantamount to treason by neglect.

In short, we can never give into an out-of-control economic ethic. We have to believe that we can and must survive without unmitigated growth. While we are led to believe that consumerism and a strong military are at the core of our culture, nothing could be further from the truth. Our new success will be measured by implementing the principles of the manifesto; nothing less will do.

As for the structure of the global republic, I have little to add to the positive ideas expressed by Ferencz, Keyes, Laszlo and others. The American model of a legislative, executive and judicial branches with checks and balances is a good place to start. But I am concerned about the recent excessive executive power in the U.S. today, for example, the President's right to wage war, influence judicial and legislative authority, ignore environmental regulation, keep secrets, and share in the largess of deep corporate pockets. This kind of power could not be legitimate in global government. I propose that the global executive branch be governed by an elected rotating council of elders.

Democracies can be inefficient, certainly more so than the authoritarian profit-driven corporations which are swallowing them up. It can take time for the will of the people to percolate outward as it becomes the laws of the land. So be it. We will need to tolerate these inefficiencies. We have no choice in that, but we could certainly innovate in finding faster ways to make the people's will known, such as frequent polls, meetings and Internet communications.

As the human slaughter of nature further escalates, we don't have much time to wait before we get moving. Yet at the same time, we cannot become too decisive initially, too efficient, in such a pluralistic process. We'll need to strike a happy medium which would ensure success while still giving the world community many opportunities to speak out. Earlier I suggested that we could have a new global constitution framed within a few years of its beginnings, which could be now. What a refreshing contrast this would be to the aggravating political transition we Americans just experienced during 2000-01. Are we up to doing this? Yes, we must. We have no choice.

Gaia is in the emergency room. "In the short term, the Earth is in need of allopathic medicine", said my colleague and friend Swami Swaroopananda at the Sivananda Yoga Ashram. In the long run, we can move to a lasting spiritual medicine based on our awakening consciousness and unity. Meanwhile, let us pray that Gaia will respond to our help in this, the eleventh hour. Meredith said: "At the very least, may we appreciate Gaia while she lasts, and may she know she is loved and is not being taken for granted." Let's savor the Earth as well as save her.

References for Chapter 7

1. Ervin Laszlo, *"Manifesto on the Spirit of Planetary Consciousness"*, The Club of Budapest, Budapest, Hungary and London, 1996.

2. Ervin Laszlo, *"On the Need for a New Spirit of Sustainability,* ibid."

3. Benjamin B. Ferencz and Ken Keyes, Jr., *PlanetHood*, Love Line Press, Coos Bay, Oregon, 1991.

8

Overcome the Insurmountable
with Compassion

"In wilderness is the preservation of the world."

-Henry David Thoreau

MY INTENTION IN writing this book has been to present a practical blueprint for transcending our destructive practices, a logical but politically incorrect quest for sustainability and for greater truth. I wanted to create a convenient text for those who choose to know what we could do to avoid planetary catastrophe. At the very least it could be a glimpse at the millennial boundary about where we are and where we could go, as a chronicle for future generations about how we either fell off the path or got back onto it in this eleventh hour. Am I being naively idealistic in this attempt? Many of you would think so. Yet we have no choice but to keep digging, to keep trying.

In sharing some basic concepts of global citizenship in this part of the book, I have learned that democracy is a highly-charged word. When combined with unchecked capitalism within a two-party system, the resulting brew creates a new elite of wealthy and ambitious leaders insensitive to the needs of the common person and life on our precious planet. The decline of the American democracy beholden to corporate hegemony echoes the failure of Soviet communism and the fall of Rome.

Some have criticized my attempts to propose a democracy at all. Citing the tyranny of the majority in our flawed system, my colleague and friend Obadiah Harris, president of the Philosophical

Research Society, has suggested the formation of a republic based on natural law. Following the Platonic ideal, leaders would be selected and
elected from a slate of culturally sensitive candidates dedicated to
the foundational principles established by the constitution of the republic.

We may even need to find a new word to describe what we mean
to do, because it is easy now to dismiss notions of democracy, republic,
socialism, communism, or the green movement because they have
become so consumed by mercentilism and economic government
that past and present examples bear no resemblance to the ideal. So we
have become cynical about discussing *any* form of world government
and may have thrown out the baby with the bathwater.

Overcoming Our Cynicism about Moving Forward.

It follows that we must step outside the existing system and
design a new one. What would be some of the first steps? In Chapter 4,
I proposed some goals for a new global government: ecological sustainability, nonviolence, social justice, ecolonomic conversion to green initiatives, and publicly supported research and development of solutions.
I also suggested some structural steps such as expanding the sciences of
social change, sustainability and consciousness, adopting simultaneous
policies, forming alliances, shifting power to local governments and
electing elders for an interim government.

So what's stopping us? What strategies can we develop to cir-
cumvent the greed, ignorance and suppression that opposes solutions?
This is the greatest mystery when looked at from a logical point of view.
History and psychology can provide some insight. At a most fundamental level, the problem is a very human one which ignores the gravity of
the situation.

First we shall need to hold our politicians more accountable.
Since they will do anything to garner votes, I am optimistic about John
Bunzl's "simultaneous policy" concept as an early solution. [1] Here is an
example of how this idea might work: Forty million Americans, myself
included, do not have health insurance. Say some of us organize and
make a vow not to vote for any politician who did not support a national health insurance plan. The 2000 U.S. Presidential election, for example,
was decided by only a few thousand votes. If Mr. Gore had supported national health insurance and Mr. Bush hadn't, the result of the elec-

tion surely would have gone to Mr. Gore. If both had, both would have benefitted from more votes. If neither had, perhaps Green Party candidate Ralph Nader would have broken through as a viable third party candidate.

Other simultaneous policies could have been imposed on the candidates. An environmental constituency, for example, might support only those candidates who would shift public subsidies from the fossil fuel and nuclear industries to the clean and renewable energy industry. Again, the 2000 election might have been decided by those "swing votes" supporting the greenest simultaneous policies (SPs). Without having to take out the time for creating viable new candidates or forming credible third parties, we could begin to re-enfranchise the 50 million cultural creatives in the U.S. and another 75 million in Europe. Of course, candidates would have to fulfill their promises, lest they lose votes from future simultaneous policies that would hold them accountable. Elections would then shift from voting for personalities to voting for policies. At the moment we are trapped by influence-peddling of lobbyists, the wealthy and media blitzes that distort reality.

Through adopting SPs we might begin to make the needed structural changes, such as making third- and fourth-party candidates viable in the U.S. European parliamentary democracies provide a good prototype for what could be done. As a first step, we Americans will need to enact meaningful electoral reform. We sometimes forget that Abraham Lincoln, considered by many historians as our greatest and most honest president, was elected by a plurality of votes in a four-party election.

We cannot underestimate the power of the individual in social movements and politics. We have few real leaders, but historically some have made a big difference such as Lincoln, Confucius, Plato, Gandhi, King and other free thinkers of the 1960s. The problem in today's world is, the best potential leaders are pushed ever further away from exposure to the public by the centralization of power through money and media. The System instead features well-financed candidates in a narrowly-confined two party matrix. Here we have a paradox. While we need world government with capable new leaders we cannot afford to have any more economic government. This is a tricky problem: we must create something new, for sure, yet we cannot create more bloated bureaucracies. Ultimately, the old systems will have to be jettisoned to make way for the new.

Where do we find our new leaders? I believe they must come from *outside* the System and be uninfluenced by money and corporations. Vested interests, wealth, and arrogance preying on the ignorance of the masses are the tyrannies from which we must escape, or fascism will surely result. Our new leaders should therefore be drafted from a pool of cultured, wise elders with a strong sense of intuition, ethics about the sanctity of life, awareness of the whole, spiritual motivation, listening ability and the capacity to lead and organize. They and their advisors would not want to seek these positions and they could play out their roles for a period of years, and not more, in service to the higher good. Let's expand the slate of candidates away from professional politicians, please! The new leaders should have proven their abilities in their relationships with family and community, and sense of vision.

Many of these individuals may have escaped or been driven from the public spotlight. In his seminal book *The Twilight of American Culture* [2], sociologist Morris Berman describes a new class which he calls the New Monastic Individuals (NMIs) that "belong to no class (and) have no membership in a hierarchy. They form a kind of 'unmonied aristocracy', free of bosses, supervision and what is typically called 'work'. They work very hard, in fact, but as they love their work and do it for its intrinsic interest, this work is not much different from play. In the context of contemporary American culture, such people are an anomaly, for they have no interest in the world of business success and mass consumerism." (pp. 135-6)

These individuals, he writes, are "disaffected Americans who feel increasingly unable to fit into this society and who also feel that the culture has to change if it is to survive." (p. 132)

"Points are reached, only to be left behind. The road to truth is always under construction; the going is the goal...An NMI doesn't participate in anything that can be labeled as an 'ism'. She might be an independent woman, but never a feminist; he might do environmental work but keep his distance from Greenpeace. For an NMI knows the historical irony of how movements start out with vibrant critical energy and wind up as new (oppressive) orthodoxies, complete with texts, heroes and slogans...the NMI is the purist embodiment of the human spirit." (p. 138)

Berman believes that these "nomads", if they were to enter political office, would give up their autonomy. "Those genuinely committed to the monastic option," he writes, "need to stay out of the public eye; they do their work quietly, and deliberately avoid media attention. Indeed, a Taoist rule of thumb might be that if the larger culture knows about it, then it's not the real thing." (p. 131)   Berman argues that NMIs will leave their historical mark by preserving a wisdom in our dark age as the monks had done in the middle ages.

On the other hand, I believe our global government could be designed to include these people. At the very least, they could quietly advise leaders to make a wise transition away from economic government towards a truly humane system free of vested interests and with openness to solutions. These are the true visionaries, the keepers of the flame. Until the press exposed them, forcing a premature ending of the relationship, the mystical anthropologist Jean Huston had advised first lady Hillary Clinton in the finest tradition of counseling leaders from an unvested and philosophical perspective.

I admit to favor the views of these monastic counterculturals, perhaps because I am rapidly becoming one myself--even if by default. This new identity has taken me decades to develop and is still unfolding.

To some readers, my mainstream credentials might appear to place me in the highest regard by those in academic, governmental, and business circles. In truth, I have left many of these institutions far behind and they have let me go as well. I have been unwittingly moving towards the kind of creative expression which often has no markets, no recognition. The glamour and glitz of sound bites and cultural varnish are fading from my life. By default, my attempts to maintain my status as entrepreneur have often ended in numerous business failures and cancelled contracts in spite of repeated attempts to penetrate the society with my works and in spite of efforts to conduct business openly, compassionately and honestly. Yet I carry on, presenting to those who will listen and dialogue with me.   Some have called me a "nomad scientist" (preferable to being a mad scientist), traveling the world and moving through institutions rather than becoming established within them. The Hindu culture encourages those of us over sixty to be wayfaring teachers and students of life.

I have become humbled in the process. The closer I come to embracing my inner-outer truth, the further from the mass culture I feel driven, yet the more joyful my creativity and spiritual practice have become. Feeding on a declining Roman Empire is not nearly as satisfying as the path of the yogi and expresser of a truth that bears little resemblance to the economic mainstream.

I hadn't expected any of this to happen. I am in this culture but no longer of it, lest I lose my way. This gradual process of liberation has been often painful. My quiet refuge in personal friendship, yoga, appreciation of nature, and creative expression and activism, free me from the shackles of the marketplace, public recognition or career advancement. One discovery I made about my severance from society came during years of retreat in the woods of Oregon while writing my previous book *Miracle in the Void*. I found myself grieving the loss of nature while being in the midst of it.   My income plummeted and my audiences shrank. I had to admit I was often unhappy and needed to do deeper work, but didn't know how to go about it, until I learned about the new discipline called ecopsychology.

Ecopsychology

My first few years as a deep ecologist had resulted in such personal grief, only recently did I become aware I was not alone in feeling this. Historian and philosopher Theodore Roszac described his own discovery of this phenomenon as follows:

"Sometime in the mid-eighties I began to realize I was burning out as a writer and speaker of environmental issues because so much of what I was presenting was relentlessly negative. The issues were legitimate, but they were taking a heavy toll on me because I was doing a lot of blaming. Burnout happens when you make an issue so impossibly large that it's difficult to see how most people could make all the changes they're asking of them quickly enough to make a difference." [3]

"I also began to realize that it was more and more difficult to connect with the people I was addressing. They were going numb on me or turning hostile. I saw this not just in my own experience but in a backlash to the environmental movement."   At last I began to understand the source of my own grief for Mother Earth and why it has been so difficult for me to share the basic facts and solutions. The challenge is psychological as well as social.

Roszac continues: "Ecopsychology seeks to find the underlying motivations for our bad environmental habits, based on the assumption that because we have an emotional bond to the planet—I call it an 'ecological unconscious'—people do want to be good environmental citizens. If they're not, there must be reasons. If you can find those reasons, you can treat them and change the behavior. That's a much more optimistic approach than simply seeing people as wicked, greedy, or hopeless and seeking to punish them for what they do."

"There is a psychological dimension to every environmental issue. These issues are not simply matters of facts and figures, not just impersonal economic forces. They are deeply personal in character, and if you haven't included that personal dimension, then you haven't included enough to solve the issue. Both of these communities have to learn from one another. Psychology needs ecology; ecology needs psychology."

Another discovery I made was my own dysfunctional behavior in light of the hopelessness about the situation. "Addiction", said Roszac, "is something I've come to focus on more and more in the work I'm doing with ecopsychologists. It is entirely possible that a large number of environmental problems that we can track with facts and figures are far more than simple moral problems. We usually think of moral problems as a direct choice between doing the right thing and doing the wrong thing. Where addiction is involved, you're dealing with people who clearly know what the right thing is, but they can't do it; they know what the wrong thing is, but they can't stop doing it." [3]

Roszac's experience and analysis provide a lucid account of why it has become so difficult for ecologists to avoid suffering and to penetrate the culture, how the environmental movement has become almost the exclusive province of lawyers, politicians, businessmen, journalists and entertainers rather than including aware scientists, visionaries and deep ecologists. It may also explain why books such as this one have been just about impossible to market. People don't want to hear about these things even though they're true! Even some of the most sensitive among us apparently would prefer to end civilization than to respond logically to bad news and move into solutions. Therefore, they (we) cover up the truth with self- and Earth-destroying addictions such as runaway consumerism, which can only deepen the despair.

In vivid contrast, those in the cultural mainstream who don't seem to care, are unaware there is any problem at all. They appear to be spared of the shame we sensitives feel. "Widespread ecological illiteracy", said Roszac, "is one of the roots of our environmental crisis. Many people simply do not understand the biological foundations of their own survival." [3].

Psychologist Sarah Conn eloquently expresses our dilemma: "The Earth hurts; it needs healing; it is speaking through us; and it speaks the loudest through the most sensitive of us. I believe that that pain wants to speak through a great many more of us. When people are unable to grieve personal losses openly and with others, they numb themselves...Many of us have learned to walk, breathe, look, and listen less, to numb our senses to both the pain and the beauty of the natural world, living so-called personal lives, suffering in what we feel are 'merely personal' ways, keeping our grief even from ourselves. Feeling empty, we then project our feelings onto others, or engage in compulsive, unsatisfactory activities that neither nourish us nor contribute to the healing of the larger context. Perhaps the currently high incidence of depression is in part a signal of our bleeding at the roots, being cut off from the natural world, no longer as able to cry at its pain or to thrill at its beauty." [4]

Combining Solutions with Compassion for All Players

Ecopsychology, then, is the missing piece for understanding and acting upon our assault on Gaia. We cannot berate ourselves and others with a guilt trip lest our paralysis deepen even more. Instead we need to move through the blame and shame and become co-creators with one another and with the Earth. The first step is to talk about it and to enter into the solutions with courage and conviction. We need to let our successes be our measure of what we can do, even if they're small at the beginning.

Above all, we need to be compassionate about what's happening, about how uniquely human our global dilemma is; about the fact that our anthropocentrism may have deeper causes yearning for a solution. We are *all* playing out roles as in a Greek or Shakespearean tragedy. On the one hand, those of us who are sensitive deep ecologists will need to heal our own grief. Otherwise, we will be paralyzed by a shame that blocks the way to personal and planetary transformation.

On the other hand, we need to understand those who espouse capitalistic competition can lead us not only to a denial of ecological truth and its biological underpinnings. The System also dictates that large corporations must grow to survive, thus seeking mergers and going wherever the labor is cheapest and environmental standards the loosest. No corporate executive in today's world would want to support the needed environmental and social programs lest they lose their business to other companies or incur bankruptcy or a hostile takeover. No politician wants to enact reform because their nation might lose business to others and thus incur growing debt and poverty. They are often bribed to oppose reform. The labor movement also needs to know that the best and most lasting jobs will come from restoring the environment, not from drilling for more oil in the Arctic or cutting down more trees. The result of all this is a short-term feeding frenzy among the rich and powerful at the expense of humanity and Earth herself.

The other threat to our survival is violence and war, and I'll try to answer the most difficult contemporary question of all: how do we deal with terrorism? My answer is quite different from the current U.S. administration's. It is very disconcerting to see that pre-emptive wars, the deployment of space weapons and possible use of nuclear weapons are even being considered at a time they should be banned for all time. Neither of these outdated and destabilizing relics of Cold War thinking could help eliminate terrorism anyway; it would encourage it and isolate America from the rest of the world.

Fascism and holocaust are lurking around the corner, and all this is unnecessary. Instead, we need to step back and ask, why have the actions of terrorists taken place? What have *we* done to help create this dangerous polarization? We must also know thine enemy and negotiate differences, or World War III could end it all.

We must also examine gaps in intelligence and the protection of our citizens and fire those who are responsible for abrogating their duty. We can choose to render humanitarian aid to innocent victims of war. We could also transform the war machine into a concerted effort to end pollution and to apply intense international pressure for finding and bringing terrorists to justice. *We could end dependence on Mideast oil by means of an Apollo program for converting to clean and renewable energy.* There are encouraging signs on cold fusion coming from the Oak Ridge National Laboratory, with peer-reviewed results being published in the March 8, 2002 issue of the prestigious journal *Science*—in spite of shrill objections by the mainstream hot fusion physics community who have yet to achieve energy "breakeven" and have spent billions of public dollars trying to do so. This apparent cold fusion breakthrough could lead to abundant, cheap and clean "free" energy through acoustic cavitation and sonoluminescence (one of the six approaches to new energy described in Appendix I).

In closing, I cannot help but express a deeper concern which we must deal with. I've done everything in my power to be objective and understated in assessing the violent politics of America in this book, to give the benefit of the doubt for those who lead, even though I don't agree with them. I cannot do so any more.

My heart recently sank with the news that the most progressive of our senators, Paul Wellstone of Minnesota, and seven others perished in a plane crash on October 25, 2002, just days before his re-election bid in a close race (Wellstone had been expanding his lead in the polls). His Republican opponent had been hand-picked by the Bush administration, who had "targeted" Wellstone above all others in the November 2002 mid-term election campaign. Wellstone had been the only Senator in a close race who voted "no" on the resolution to give Bush carte blanche to attack Iraq. The tottering balance of power in the Senate is crucial to the administration's desire to wage war, rape the environment, flaunt laws and treaties, protect the wealthy and ignore the rest of us in America and worldwide. The elections of course turned out giving the President the majorities he needed, including the Minnesota race, so we can watch reality unfold in stark relief to what it could be, with gobs of money and possible assassination (shhh) as contributing factors. The media calls it "Bush's popularity" and boasts about how much money the Republicans spent to take control of Congress.

Having been a pilot myself, I am very suspicious of the cause of the crash, as I had about Missouri Democratic Senatorial candidate Mel Carnahan's uncannily similar fate just before the 2000 election and the unpunished anthrax attacks of autumn 2001 which had targeted Democratic Senate leaders Tom Daschle and Patrick Leahy. Whatever the cause of Wellstone's "accident" or the others, whatever our interpretation we might have about the administration's overt and covert actions, the Wellstone tragedy is for me a poignant trigger to wake up even more before it's too late—reminiscent of the murders of the Kennedys.

In my 62 years here I have never seen such a radical move away from our ideals in an administration—even during our disastrous involvement in Vietnam which had led me in May 1970 into the Nixon White House as a protester and then appearing on the CBS Evening News anchored by Walter Cronkite. (The most recent anti-war protests in Washington and elsewhere were barely covered by the news). Cronkite in a recent speech warned that the Bush administration was inviting World War III in its belligerent actions. This pre-election war vendetta has been called by Michael Moore "weapons of mass distraction." Unfortunately the strategy worked and the Democrats rolled over.

In 2004, we will have to remove our current leaders from

power, using whatever nonviolent and democratic means possible. They're as corruptly big business-controlled as any regime in the U.S., warned about by so many presidents who had a deeper sense of the purpose of our democracy referred to in this book. This is a call to action, to massive protests and civil disobedience. In protesting the American aggression against Mexico in 1848 and spending a night in jail for doing so, Henry David Thoreau wrote an essay *On the Duty of Civil Disobedience*: "Cast your whole vote, not a strip of paper merely, but your whole influence. A minority is powerless when it conforms to the majority; it is not even a minority then; but it is irresistible when it clogs by its whole weight."

The challenge and adventure are before us. I hold the vision that we can join the forces of nonviolence, sustainability, human rights and greater knowledge to forge a new world. We need to find ways besides competition and profit to do our work. And we who want to heal ourselves from our denials and guilt have an opportunity as never before to step outside the System and re-inherit the Earth.

The divergence between spiritual reality and imperiled world has never been greater. The opportunity to transcend this dilemma has never been greater. Combining our focused intelligence and compassion with the peaceful will of God can do the job, so let's get on with it!

References for Chapter 8

1. John M. Bunzl, *The Simultaneous Policy*, New European Publications, London, 2001.

2. Morris Berman, *The Twilight of American Culture*, W.W. Norton, New York, 2000.

3. Theodore Roszac, *The Voice of the Earth*, Touchstone, 1995, and interviews with D. Patrick Miller fromThe Sun and with Suzi Gablik in *Common Boundary*, March/April 1995; also *Discussion Course on Deep Ecology*, Northwest Earth Institute, Portland, OR.

4. Sarah A. Conn from *Ecopsychology* by Theodore Roszak, Mary E. Gomes, and Allan D. Kanner, 1995; also in *Discussion Course on Deep Ecology*.

APPENDIX I

Review Article on New Energy

MIRACLE IN THE VOID:
THE NEW ENERGY REVOLUTION

by Dr. Brian O'Leary and Stephen Kaplan

This article first appeared in the August 1999 *Review of the Scientific and Medical Network*, United Kingdom; updated versions can be found at www.spiritofmaat.com

"Ere many generations pass, our machinery will be driven by power obtainable at any point in the universe. . .it is a mere question of time when men will succeed in attaching their machinery to the very wheelwork of nature."
- Nikola Tesla

Imagine a world with abundant, compact, inexpensive, clean energy. Global warming has been reversed. The air is clean. Power and gas stations have been put to other uses, and the unsightly grid system has been dismantled and recycled. An unusual type of energy makes all this possible. It does not come from the sun, wind, rain or tides, nor is it a fossil or nuclear fuel taken from the ground or chemically synthesized. In some cases the source appears to be everywhere in space-time, invisible and infinite. In others, nuclear transmutations seem to be miraculously taking place at room temperature accompanying the release of energy.

Incredible as this scenario appears, we shall see in this article that laboratories around the world are repeatedly tapping into this abundant energy. Some leading theoretical physicists are beginning to understand why and how this is possible. Several companies are in the beginning stages of bringing workable devices to market that clearly produce more energy than what is needed to run them (so-called overunity devices).

Research Breakthroughs and Commercial Developments

According to knowledgeable observers, we should focus our attention on six new technologies: *(1) plasma-type devices; (2) solid-state electromagnetic devices; (3) hydrogen gas cells; (4) super motors based on super magnets; (5) cold fusion or new hydrogen energy (the Japanese name for cold fusion); (6) hydrosonic or cavitation devices.*

Plasma-type devices

A variety of plasma-type devices have been patented and are being developed with private funding for applications ranging from computer chips to power plants. One of the most promising of these devices is called the XS NRG PAGD (Pulsed Abnormal Glow Discharge) reactor. Created by inventors Dr. Paulo and Alexandra Correa at Labofex in Canada, the system is well on its way to commercialization for electric power generation in vehicles and in homes, solidly protected by U.S and foreign patents (U.S. Patent 5,449,989).

The reactor's self-oscillating electrical plasma discharge produces electrical energy directly, with no intermediate thermal conversion step. The electricity produced by the Correas' device is effectively free. The Correas write conservatively: ". . .The cost of kWh (kilowatt hours) produced by this technology is estimated to be more than ten times cheaper than what is presently available from any other energy source."

Ken Shoulders' "charge cluster" device is another promising plasma technology. Shoulders worked for a number of years as a researcher for MIT, Stanford Research Institute and private corporations. In the 1960s he helped to develop much of today's microcircuit technology. His high-density charge cluster device produces more than thirty times greater output than input energy.

Produced by a short pulse of electric potential, a typical one-micron charge cluster is a tightly packed group of about 100 billion electrons which have broken free from their nuclei and have come together into remarkably stable ring-shaped units that look like tiny donuts. At first glance, they seem to violate a principle of physics that asserts that electrical charges, either positive or negative, repel each other. Shoulders' 1991 patent for "energy conversion using high charge density" was the first successful patent to claim significantly more output than input in a device that could be a practical source of decentralized electrical energy (U.S. Patent 5,018,180). Because charge cluster technology works without the need for magnetic fields or low temperatures, it could well be one of the first new energy devices to be commercialized.

Shoulders' basic process may also be valuable for the remediation of nuclear waste. By bombarding radioactive nuclei with charge clusters, the induced nuclear reactions (primarily fissioning of the heavier elements) result in a reduction of harmful radiation. Laboratory experiments show a dramatic transmutation of radioactive thorium into smaller-mass elements with the marked reduction of the naturally radioactive thorium. With proper engineering complete eradication may become possible.

Solid-state devices

A pioneering solid-state technology is Wingate Lambertson's World into Neutrinos (WIN) process. Dr. Lambertson has conducted materials research and development for such organizations as U.S. Steel, the Universities of Toledo and Rutgers, Argonne National Laboratory, the Carborundum Company and Spindletop. He has been doing independent research over the past two decades on a a solid-state device which he believes can provide a practical source of power through the harnessing of zero-point energy.

Lambertson's electron dam (E-dam) is made out of Cermet, a highly advanced heat-resistant ceramic and metal composite. An accelerated electrical charge sends a stream of electrons into the E-dam, and the electrons become stored much as a conventional dam stores water. When the electrons are released, they gain energy from the zero-point energy present in the E-dam. After they flow into the unit to be powered, they move into another E-dam for recycling.

Lambertson changed his cermet chemistry and E-dam design

when he learned that an unexpected chemical reaction was taking place. A different combination of materials and composite design appears to stabilize the process, and a yield of 145 percent was achieved in tests conducted in 1998. Since that time an induction effect has become a major problem which severely inhibits charge acceleration and yield. The present direction of his research is towards reducing induction in his E-dam using two different complementary approaches. It appears that these approaches will solve his remaining major problem. His highest yield using these approaches in June 1999 was 109 percent.

Lambertson is confident that he will achieve higher yields with further experimentation, probably as high as 200 per cent, the level needed for commercial viability. He is currently exploring future production with interested manufacturers. Lambertson has a strong interest in providing new solutions for the energy needs of developing nations.

Highly regarded Canadian inventor John Hutchison has developed a solid state crystal energy converter made out of very common materials which is an electrical power source he claims behaves like a battery and never runs down. This small, self-running power source, which typically puts out DC power amounting to one or two volts, has produced up to six watts of power, and he believes it could be engineered to replace batteries and other power needs.

Hydrogen gas cell

Dr. Randell Mills of BlackLight Power, Inc. has come up with a simple device he believes turns hydrogen into a clean and limitless source of power. In his lab, Mills puts in a small amount of hydrogen gas into a vacuum-sealed, three-quart stainless steel can, adds a few tablespoons of a common chemical compound, turns up the heat to about 250 degrees centigrade and reportedly creates ten to twenty times more energy than he put into the process.

According to Mills, the secret to BlackLight Power comes from shrinking or collapsing the size of the hydrogen atom from its natural ground state to a lower energy state. As the hydrogen nucleus collapses, the electron gives off heat energy, and the smaller the nucleus gets, the more heat the electron gives off. Once started, the reaction can sustain itself, as the hydrogen atoms collapse smaller and smaller, releasing increasing amounts of energy with each collapsing "transition". Of course, we do not yet know whether Mills' theory or some other will eventually be accepted by science as the explanation for this process. The important point is that the experimental results show that it works.

An impressive feature of the BlackLight Power gas cell is its simplicity. If successful, the process is also attractive because it is safe, lacks harmful byproducts or emissions, and does not require the use of exotic or scarce materials. The only waste products are collapsed hydrogen atoms, or what Mills calls hydrinos, which have characteristics similar to helium, being inert and lighter than air.

Because hydrogen is the most abundant element both on Earth and throughout the universe, the abundance of energy available through this process is difficult even to imagine. For example, a cup of water contains enough hydrogen to produce over 3,000 kilowatt-hours of heat if you collapse the hydrogen atoms to one-twentieth of their normal size, or enough to provide all the energy needed for an average home—space and water heating plus electricity—for a month.

PacifiCorp, an Oregon-based utility holding firm, has demonstrated its belief in Mills' work by investing more than one million dollars in the company, and other corporations appear likely to follow their lead. With this kind of support, Mills is confident his company will soon complete a BlackLight Power cell that will produce a full kilowatt of heat. Once this is accomplished, it can quickly bring to market a commercially viable energy cell.

Magnetic Motors

Based on physicist Michael Faraday's observation in the 1830s that anomalous electricity can come off a rotating disk containing magnets, a number of inventors have created motors that they believe have produced over-unity power in public demonstrations. One of us (O'Leary) watched several such demonstrations, as reported in the book *Miracle in the Void*. According to their inventors, some of these new devices produce four to six times as much mechanical energy as input electrical energy. In other words, once a machine achieves a certain threshold of revolutions per minute, it supposedly can be unplugged and serve as a free-running generator of electricity.

A number of magnetic motors have been evaluated by Magnetic Power, Inc., including devices created by the following inventors: Takahashi, Johnson, Tobias, Adams, Yassir, Werjefelt, Kawai, Sweet, Muller and Newman. They have also tested a device called the Magnetic

Wankel from the Japanese firm, Kure Teko. None of these devices has yet proven to function over-unity under load.

According to Magnetic Power, there are two eminent scientists whose work on magnetic devices may some day bear fruit: Dr. Paramahamsa Tewari, Chief Project Engineer for India's Kaiga nuclear power plant construction program, who is developing what he claims is a space energy motor/generator with support from the Indian government; and Dr. Harold Aspden in Great Britain, former patent counsel for IBM Europe, who is working on his version of a practical over-unity magnetic motor.

Many of the devices developed by inventors have proven to be merely magnetic flywheels and are therefore in fact only a type of mechanical energy storage device. As our still incomplete understanding of magnetism continues to grow, it is possible that one day a design based on a new source of energy will prove practical. Roots, a subsidiary of Magnetic Power, is developing room-temperature ultraconductors. Made of highly conductive polymer materials, this technology may make possible the creation of very powerful, lightweight motors and generators without the use of iron or copper.

Cold fusion

On March 23, 1989 Drs. Martin Fleischmann and Stanley Pons at the University of Utah startled the world with their claim that they were getting excess thermal energy far beyond electrical input energy from an electrochemical cell with a palladium cathode and a heavy water electrolyte. The cell output, they said, was substantially more than could be explained by any chemical reactions.

Considerable controversy has surrounded cold fusion ever since the first experiments. In 1989, a Department of Energy research panel, dominated by hot-fusion advocates, proclaimed that no government funding should be invested in such a questionable area of research. Since then, following the lead of the scientific establishment, the media have generally either criticized or ignored cold-fusion researchers. However, the evidence for the validity of cold fusion is growing stronger and stronger daily. The peer-reviewed published literature provides overwhelming support for both the nuclear-scale excess heat and nuclear changes in what were supposed to be exclusively chemically active systems. There have been reports of transmutations of heavy elements in various cold-fusion experiments—both in ordinary water and in heavy water systems— potassium changed to calcium, rubidium to strontium, and palladium to silver, rhodium, ruthenium, etc.

The Pons-Fleischmann process has been declared valid by Dr. Georges Lonchampt, one of the top members of the French Atomic Energy Agency. Researchers at Los Alamos National Laboratory and the U.S. Navy's China Lake research lab have conducted successful cold-fusion experiments. Finally, the U.S. Army has reviewed the pioneering cold-fusion research of Dr. John Dash, a metallurgist from Portland State University in Oregon, and decided to fund his work for three years.

Stanley Pons and Martin Fleischmann, despite overwhelming criticism heaped on them by US scientists and scoffing by American media, continued their work in France and England. With funding from a Toyota foundation, they claim to have have made excellent strides in the development of the heavy-water, palladium cathode, electrochemical cells. Currently, months of continuous testing of cells operating at boiling temperature has produced data showing thermal output of twice the input electrical power all from a tiny palladium cathode.

Clean Energy Technologies (CETI) is marketing licenses for a power cell invented by Dr. James Patterson, a scientist with a distinguished record of achievement.   News of his device spread widely when it was discussed on two *ABC* shows, *Nightline* and *Good Morning America.*

The Power Cell has produced kilowatt levels of thermal energy at more than twenty times the input electrical energy. It has been independently tested and replicated by several universities, utilities and corporate research laboratories. Motorola has tested a number of cells and found that with at least one cell they were able to turn the input electrolysis power off, walk away, and have the output power of twenty watts (thermal) continue for at least a half a day.   On June 11, 1997, CETI announced on *Good Morning America* that a prototype hot water heater is scheduled to be mass-produced within two to four years.

Perhaps the most astonishing finding from cold-fusion research is the apparent observation of radioactivity reduction in the process! CETI, one of the first cold fusion companies, recently announced it had been awarded a US patent on an electrolytic process for reducing the radioactivity of thorium and uranium. The company claims its process can reduce the radioactivity of radioactive materials by over 90 per cent in periods less than 24 hours, compressing into hours what nature takes billions of years to do. A demonstration of this reportedly successful process was included in the same Good Morning America story which described Patterson's prototype water heater.

Dr. Norm Olson, a Department of Energy researcher based at its Hanford nuclear facility, was interviewed by ABC and indicated an interest in exploring Patterson's process. He later tested the CETI power cell and found that it did indeed reduce the radioactivity of uranium and thorium. He cautions, however, that much more basic research needs to be done before this or any other process can be developed into a workable technology for dealing with nuclear waste.

There are at least three other groups who also claim to be able to reduce radioactivity by other methods, which have yet to be awarded patents. One of them, the Cincinnati Group, is marketing to scientists a kit which demonstrates their transmutation process. If purchasers follow their suggested protocol and do not get the claimed results, their money will be refunded.

Very much worth watching is a cold-fusion invention announced at the most International Conference on Cold Fusion (ICCF-7), held in Vancouver, B.C. in April 1998. It is the "catalytic fusion" process of MIT-educated chemical engineer Dr. Les Case. Based on more than six years of painstaking research, Dr. Case's discovery appears to be a nearly optimal embodiment of the original Fleischmann-Pons process. A pre-treated activated carbon catalyst with 0.5% to 1.0% palladium or other catalytic metal content apparently can catalyze the fusion reaction of heavy hydrogen (deuterium) gas to helium at elevated temperature (150 to 250 C). In tests conducted by Dr. Eugene Mallove of *Infinite Energy* magazine, it was confirmed that the Case process achieved a persisting excess temperature that climbed to 13.2 C above the baseline temperature of 178.1 C, which represents approximately 7.5 watts excess power. It exhibits the heat-after-death phenomenon that many researchers have reported coming from cold-fusion cells: heat production with no input power after the reaction is triggered.

If replication holds up, it may be impossible for anyone to deny Dr. Case's process, especially once he has made a device that can self-sustain—that is, employ no electrical heater power. Mallove believes that the process looks well-positioned to be a simple commercial power-generating technology in small, distributed units, as well as in large power plants.

Hydrosonic or cavitation devices

James Griggs' Hydrosonic Pump is already being sold to customers, regularly providing them with over-unity energy. An energy efficiency consultant from Georgia, Griggs invented the pump as a result of his curiosity about a common phenomenon called water hammer or cavitation. Griggs noticed that heat emanated from fluids which flow quickly through the pipes of a boiler, causing water pressure to drop in part of the pipe. Bubbles formed in the low-pressure areas collapse when carried to areas of higher pressure. The resulting shock waves collide inside the pipe, bringing about the water hammer effect.

Griggs' pump is made of a cylindrical rotor that fits closely within a steel case. When the rotor spins, water is forced through the shallow space between the rotor and the case. The resulting acceleration and turbulence created in the gap somehow heats the water and creates steam. In 1988, a testing expert found that the heat energy put out by the hydrosonic pump was 10 to 30% higher than energy used to turn the rotor.

In 1990, Griggs started Hydrodynamics, Inc. He and his partner have invested over a million dollars in the business. The units they are selling are not only more efficient than standard boilers but they also require less maintenance. They are self-cleaning and eliminate the problem of mineral build-up that reduces the efficiency of standard boilers. Georgia Power and the civil engineering department at Georgia Institute of Technology are currently conducting studies of the pump.

A new cavitation device similar to the Griggs machine is now available for testing, scientific investigation and purchase by research laboratories. This is the "Kinetic Furnace" of Kinetic Heating Systems, Inc. of Cumming, Georgia. Jointly invented by Eugene Perkins and Ralph E. Pope, the furnace is a heat-producing rotary cavitation device for which the inventors have been granted four United States patents, the most recent one in 1994. Numerous independent companies and testing agencies have found the same over-unity performance: Coefficient of Performance or C.O.P. (the ratio of output to input power) in the range 1.2 to as high as 7.0, with most typical operation in the range 1.5 to 2.0. Dr. Mallove and Jed Rothwell of Infinite Energy recently confirmed the excess heat in a preliminary on-site test.

The reactions responsible for the excess energy in the Perkins-Pope device may be novel nuclear reactions or the tapping of energy reservoirs that some have referred to as new hydrogen energy states or zero-point energy. There is no possibility, according to Dr. Mallove, that the device can be explained by chemical energy or "storage energy".

The Kinetic Furnace represents a technology that will have application in water and air heating, and perhaps in self-standing electric power production and rotary mechanical power production. One of the largest commercial hot water heater manufacturers in the world, State Industries of Tennessee, has been calling regularly to check on the inventors' progress. That company, and no doubt others, are taking a keen interest in the near-term prospect of equipping their commercial water heating systems with devices that could save the consumer 30% or more of their electric utility bills.

First Gate Energies (formerly known as E-Quest Sciences) has created devices that use ultrasound induced cavitation to produce large amounts (hundreds of watts) of anomalous excess energy. Experimental devices also produce helium and in some experiments, tritium. Yet the process does not produce any significant amount of penetrating radiation or nuclear waste—well below safe exposure levels.

There is no consensus regarding how this process works. According to one hypothesis, a myriad of collapsing bubbles form during multi-bubble cavitation produced by intense ultrasound, and the individual bubbles act like micro-accelerators injecting deuterons and other ions into nearby solid lattices. Under the influence of the lattice and with other stimulation, nuclear reactions involving deuterons and other nuclei are initiated and controlled.

First Gate has successfully demonstrated their devices and methods at Los Alamos National Laboratory and at SRI International. It is currently seeking strategic alliances with companies which can help them develop commercial products such as hot water heaters and space heaters. There are also the recent breakthroughs in acoustic cavitation at Oak Ridge National Laboratories referred to in Chapter 8.

The Physics of Zero-Point Energy

As impressive as the experimental evidence may be, most scientists do not seem to understand how these free-energy devices could work, since they seem to violate the laws of thermodynamics. Only when the new processes are understood properly will it be seen that they behave lawfully. We see a growing consensus among some physicists and systems engineers concerning the changes in theoretical models that will be required to account for the growing number of experiments pointing to this phenomenon.

Some physicists have hypothesized that an all-pervasive electromagnetic energy field must be present for electrons and other particles to be continually vibrating as they radiate away energy. They assert that this "vacuum energy" (sometimes called the Casimir Effect) is a fundamental property of nature, a conclusion based on experiments and observations in quantum mechanics.

Physicists Harold Puthoff, Bernhard Haisch and Alfonso Rueda have recently published papers on this in the prestigious peer-reviewed journal *Physical Review A* and in the highly regarded periodicals *The Sciences* and *Mercury*. In these articles they suggest that if an electromagnetic charge is accelerated in the zero-point field, there may exist conditions which allow for either the extraction of energy from the field, or dissolution of energy into the field. This accelerative force might then break the monotony of homogeneity and isotropy that has held us back from perceiving the vast benefits that lie within the vacuum of space. In fact, the rotary motions of fluids and magnets, as well as the resonant vibrations inside solid-state devices, appear to provide the needed accelerative forces to tap the zero-point field.

One of us (O'Leary) has also looked at the possibility that the zero-point field concept might be related to the mysterious interactions found in quantum mechanics and psychic phenomena. In combination, these anomalies of materialistic science might lead us to look toward creation of a new science of consciousness.

The Environmental Imperative

During the 1970s, many of us were first exposed to some of growing global energy and environmental problems. The OPEC Cartel, gas station lines, smog, oil spills, oil wars, nuclear accidents, nuclear proliferation and waste disposal were in the headlines and began to become known by the public (i.e., U.S. citizens) and to stimulate needed policy changes.

The global situation is even more critical now than it was in the 1970s. Global warming from greenhouse gases coming mainly from the burning of fossil fuels may be the most ominous sign of possibly irreversible damage. A growing consensus among mainstream climatologists is that the unprecedented persistence of El Nino conditions and resulting local weather extremes can be directly traced to overuse of oil and coal for electricity, heating and transportation. Incremental reforms such as emission controls, energy conservation and turning to renewable sources, while helpful, are clearly insufficient to stop, much less reverse, current trends.

Widespread loss of life and deteriorating health throughout the world are caused by both the pollution and the weather shifts. Moreover, these conditions have led to an acceleration of airborne diseases that have already killed hundreds in the affected areas. Public health experts are warning that this may be only the beginning of widespread plagues. Some scientists have warned that global warming is also accelerating ozone depletion, an even more serious threat to life on Earth.

Trends in energy use make the picture look more menacing. Since the time the energy crisis first erupted, we have tripled oil consumption and doubled electricity use worldwide. *Fortune* magazine has noted that if the per capita energy consumption of China and India rises to that of South Korea (which we believe is a realistic scenario), and the Chinese and Indian populations increase at currently projected rates, these two countries alone will need a total of 119 million barrels of oil a day, almost double the world's entire demand today. The burning of fossil fuels by all nations at projected levels not far in the future might simply make our planet uninhabitable.

The Transition to a Free Energy Economy

Free-energy technology holds immense promise for alleviating or eliminating entirely the threat posed by fossil fuel and nuclear pollution. It will also allow all areas of the world, including both developed and developing economies, to meet increasing needs for energy without bringing about environmental disaster.

The rate of invention and communication in the new energy field is accelerating rapidly. Thousands of experiments are being conducted around the world, some by inventors working alone and others by collaborative efforts. Granted, we are largely in the research phase of the research and development cycle, with no guarantees that any particular commercial system will become available within a certain time frame. Granted, too, there have been and will be many false starts and exaggerated claims. However, given the significant initiatives in cold fusion and other new-energy technologies in the U.S., Japan and elsewhere, there is little doubt that workable free-energy devices will be available soon.

Japan may be first country to achieve a breakthrough to commercially available technology. With minimal domestic energy supply, Japan has little to lose and much to gain from developing new energy sources. There are more institutional barriers to, and less incentive for, the development of free energy in the United States. Yet work here is widespread, and projects are beginning to attract funding from a wide variety of investors, including utilities.

The revolution in personal computing is helping to speed up the free-energy revolution. A number of dynamic free-energy discussion groups can be found on the Internet, characterized by spirited discussion of emerging issues and down-to-earth exchange of research data and suggestions. Because of the distinctive openness of Internet communication, cooperation among participating inventors and scientists almost certainly guarantees that potentially workable theories and processes will be widely replicated and tested. Furthermore, no central authority will be able to squelch progress, as it might have been able to do in the past when individual inventors almost always worked in isolation.

We are in the midst of a scientific revolution of unprecedented magnitude. The necessity to overhaul our currently bankrupt energy systems to save the environment is an idea whose time has come. As we bring into existence new forms of power, we will bring about social and economic change unparalleled in human history. The situation we are in demands fresh perspectives and inspired leadership. Can we successfully master the challenges this transformation presents? We believe we must and can do so, but only if we start now the kind of democratic discussion and action this momentous shift requires.

A Bibliography and Guide to Resources

What follows is a listing of the sources of information on the free-energy revolution. Included are major books and articles, persons and organizations which are authorities in the field, and selected web sites.

In preparing this resource, we have gained valuable knowledge and insight from a few knowledgeable people who keep close watch over developments in the new-energy field, including Dr. Hal Fox, editor of *New Energy News* and the *Journal of New Energy*, Thomas Valone of the *Integrity Research Institute*, Dr. Eugene Mallove of *Infinite Energy* magazine, Bruce Miland of *Electrifying Times*, and Jeane Manning, author of *The Coming Energy Revolution*.

Re-Inheriting the Earth

Books

Bearden, Thomas, *The New Tesla Electromagnetics and the Secrets of Electrical Free Energy*, Tesla Book Company, Chula Vista CA, 1990.

Cheney, Margaret, *Tesla: Man Out of Time*, Dell Publishing, New York, 1981.

Childress, David Hatcher, *The Free Energy Device Handbook*, Adventures Unlimited Press, Kempton IL, 1994.

Davidson, John, *The Secret of the Creative Vacuum*, C.W. Daniel Co. Ltd., Essex, England, 1989.

Davidson, John, *Subtle Energy*, C.W. Daniel Co. Ltd., Essex, England, 1987.

Eisen, Jonathan, editor, *Suppressed Inventions and Other Discoveries*, Auckland Institute of Technology, Auckland, New Zealand, 1994.

Fox, Hal, *Space Energy Impact in the 21st Century*. Fusion Information Center, Salt Lake City, UT.

Fox, Hal, *Cold Fusion Impact in the Enhanced Energy Age*, Fusion Information Center, Salt Lake City, UT, 1992; Bibliography on computer disk available in English, Russian and Spanish.

Kelly, Don, *The Manual of Free Energy Devices and Systems*, Cadake Industries Inc., Clayton, GA, 1987.

King, Moray B., *Tapping the Zero-Point Energy*, Paraclete Publishing, P.O. Box 859, Provo, UT 84603, 1989.

Lindemann, Peter A. *A History of Free Energy Discoveries*. Bayside, CA: Borderland Sciences Research Foundation, 1986.

Mallove, Eugene F., *Fire From Ice: Searching for the Truth Behind the Cold Fusion Furor*, John Wiley & Sons, Inc., New York, 1991.

Manning, Jeane, *The Coming Energy Revolution*, Avery Books, New York, 1995.
The best general journalistic piece on free energy developments.

Mizuno, Dr. Tadahiko, *Nuclear Transmutation: The Reality of Cold Fusion*. Cold Fusion Technology, Inc., P.O. Box 2816, Concord, NH 03302-2816 - An account of one scientist's experience on the frontiers of new-energy research.

Moray, T. Henry and John Moray, *The Sea of Energy*, Cosray Research Institute, P.O. Box 651045, Salt Lake City UT 84165-1045, 1978.

O'Leary, Brian, *Miracle in the Void*, Kamapua'a Press, Hawaii and Colorado, 1996.

Seifert, Mark, *Wizard: The Life of Nikola Tesla.*

Talbot, Michael, *The Holographic Universe*, HarperCollins, New York, 1991.

Tewari, Paramahamsa, *Beyond Matter*, Printwell Publications, Lekh Raj Nagar, Aligarh-202001, India, 1984.

Valone, Thomas, *Electrogravitics Systems*, Integrity Research Institute, Washington DC, 1994.

Articles

Cole, Daniel C., and Harold Puthoff. "Extracting Energy and Heat from the Vacuum". *Physical Review E* Vol. 48, No. 2 (August 1993): 1562-1565.

Haisch, Bernhard, Alfonso Rueda and Harold E. Puthoff. "Beyond E=MC^2", *The Sciences* (November/December 1994) and Physical Review A (February 1994).

Hathaway, George. "The Hutchison Effect" *Electric Spacecraft Journal* Vol. 1 No. 4 (1991): 6-12.

Kestenbaum, David. "Cold Fusion: Science or Religion?" *R&D Magazine,* Vol. 39 No.4 (April 1997): 51-56.

Lambertson, Wingate. "History and Status of the WIN Process" In Proceedings of the International Symposium on New Energy, in Denver, May 12-14, 1994. Fort Collins, CO: Rocky Mountain Research Institute, 1994, 283-288.

Lindemann, Peter A. "Thermodynamics and Free Energy." *Borderlands* Vol. L No. 3 (Fall 1994): 6-10.

Mallove, Eugene F. "Cold Fusion: The 'Miracle' is No Mistake." *Analog,* July/August 1997: 53-73.

Puthoff, Harold. "Quantum Fluctuations of Empty Space: A New Rosetta Stone of Physics?" *Frontier Perspectives* Vol. 2 No. 2 (Fall/Winter 1991): 19-23.

Tesla, Nikola. "The Problem of Increasing Human Energy." *The Century Illustrated Monthly Magazine,* June 1909, 210.

Videotapes

Free Energy: The Race to Zero Point: A comprehensive 110-minute broadcast-quality documentary featuring the most promising devices, processes and theories from brilliant visionary scientists and the most persistent independent inventors on the planet. Hosted by Bill Jenkins, written and directed by Christopher Toussaint and produced by Harry Deligter. Distributed by Lightworks Audio & Video. Can be ordered from Fusion Information Center, P.O. Box 58639, Salt Lake City, Utah 84158

Cold Fusion and New Energy Briefing: Recent Breakthroughs, History, Science and Technology - a two-and-a-half hour videotaped lecture and seminar by Dr. Eugene F. Mallove, editor of *Infinite Energy.*

Order from Cold Fusion Technology, Inc., P.O. Box 2816, Concord, N.H. 03302-2816.

Cold Fusion: Fire from Water - a fast-paced documentary about what has happened to the Cold Fusion discovery in the years since the Utah announcement. The most recent and comprehensive overview of the work being carried out in laboratories and companies. Order from Cold Fusion Technology, Inc., P.O. Box 2816, Concord, NH 03302-2816.

Brian O'Leary, *Miracle in the Void: Free Energy, New Science, Consiousness and the Earth.* A two-hour lecture summarizing much of what is in this article. Order from Dr. O'Leary, P.O. Box 2003, Nevada City, CA 95959

Information Sources

A list of information sources (listed alphabetically) that are expressing interest in cold fusion or other enhanced energy devices.

Periodicals and Journals

Electrifying Times Newsletter.
Published by Bruce Meland, 63600 Deschutes Road, Bend, OR 97701. Phone 503-388-1908; Fax 503-382-0384; E-Mail 102331.2166@compuserve.com

New Energy News (NEN) INE Newsletter
The monthly newsletter of the Institute for New Energy. Salt Lake City, UT. TEL 801-583-6232, FAX 801-583-2963. Email to ine@padrak.com Web Site at http://www.padrak.com/ine/

Infinite Energy magazine - a comprehensive publication that covers R&D in cold fusion and new energy technology, the "Scientific American" of the field. Published six times a year by Cold Fusion Technology, Inc., P.O. Box 2816, Concord, New Hampshire 03302-2816. Tel. 603-228-4516; Fax: 603-224-5975. Email: editor@infinite-energy.com Web site: http://www.infinite-energy.com

Planetary Association for Clean Energy (PACE) Newsletter
Quarterly newsletter, edited by Dr. Andrew Michrowski, 100
BronsonAvenue, No. 1001, Ottawa, Ontario K1R 6G6, Canada, TEL
613-236-6265, FAX 613-235-5876.. Now available: Clean Energy
Review, a technical and scientific discussion prepared for the
Canadian Environmental Assessment Agency's panel reviewing
nuclear fuel wastes disposal. Discusses transmutation as one possible solution. $5.00 U.S. and Canadian, $7.50 other countries.

Space Energy Journal
Quarterly newsletter/magazine edited by Jim Kettner and Don
Kelly, P.O. Box 1136, Clearwater, FL 34617-1136.

Web Sites

Aether Science: Home Page of Harold Aspden
Ourworld.compuserve.com/homepages/H_ASPDEN

Bearden, Thomas: Authored Files
http://www.newphys.se/elektromagnum/physics/Bearden/

Bearden, Thomas: Virtual Times: Authored Files
http://www.hsv.com/writers/bearden/tommenu.htm

Bedini, John: Collection of Free Energy Machines
rand.nidlink.com/~john1/

BlackLight Power, Inc. Homepage
http://www.blacklightpower.com

Explore Publications (and Explore Magazine)
http://www.electriciti.com/explore/

Free Energy, Anti-Gravity & Quantum Physics: Leading Edge
http://www.trufax.org/menu/energy.html

Hot and Cold Running Fusion
http://www.skypoint.com/subscribers/jlogajan

Infinite Energy Magazine
www.infinite-energy.com

Institute for New Energy (INE)
http://www.padrak.com/ine/   Contains many important scientific papers and current reports on all areas of research. Email to
ine@padrak.com. INE, P.O. Box 58639, Salt Lake City, UT 84158-
8639. TEL 801-583-6232, FAX 801-583-2963. Features "New Energy
News", the monthly newsletter for the INE, highlighting the
research and development in the worldwide new-energy arena.
Edited by Hal Fox

Homepage by Stefan Hartmann in Germany
http://www.overunity.de

KeelyNet BBS
Science-and-health oriented information exchange that specializes
in nonstandard research, much of it on new energy. Jerry Decker,
214-324-3501. Web site at http://www.keelynet.com; e-
mailto:jdecker@keelynet.com

Brian O'Leary
http://www.independence.net/oleary/

Planetary Association for Clean Energy, Inc. (P.A.C.E.)
http://www.energie.keng.de/~pace/

Quest of Overunity, by Jean-Louis Naudin
http://www.members.aol.com/Jnaudin509/

Weird Science, Anomalous Physics and Tesla Society
http://www.eskimo.com/~billb/weird.html

Zero Point Energy and the New Physics
http://www.dnai.com/~zap/

Keynote Address: Clean Energy Now!

Speech to the Forum on Converting to a Hydrogen Economy

Ft. Collins, Colorado, Sept. 22, 2000
by Brian O'Leary, Ph.D.

I thank you for inviting me to share my insights on the world energy situation and its solutions. I come from an eclectic background in physics, aerospace engineering, and most recently, ecolonomics and politics.

First, it is clear to me that humanity is on a suicidal course towards global disaster at the rate we are burning fossil fuels. The evidence is now obvious: the consensus among climatologists is that global warming and drastic climate change now being wreaked upon us are human-caused, the collective action of the burning of fossil fuels in over a billion vehicles and in dirty power plants worldwide. Air pollution alone from these causes has created more death and illness than any second-hand smoke could ever do. I am well into a book *Re-Inheriting the Earth*, which documents these crimes against ourselves and our natural environment, and so I must place my political remarks first, with all due apologies to nonprofit structures who are not supposed to talk about such things. Yet not to do so would be massive denial of our truth. So my remarks will be political first, then technical.

That we could possibly vote in as U.S. president and vice president (George W. Bush and Dick Cheney) oilmen who believe in voluntary emission controls (and who seem to have no idea of what we are doing to our atmosphere and entire biosphere), is like asking for the foxes to guard the chicken coop. And yet our current vice president (Al Gore) has proven time and time again to have been bought out by oil companies and other polluters, and so has been silenced on the environment as a politically incorrect issue. Whatever happened to his passionate book *Earth in the Balance*, which now reads more like a treatise about a historical curiosity before he came home triumphantly from the Kyoto accords, which are too little too late and have no American support anyway? Yet under his watch with Mr. Clinton, we witness the onset of the most gas-guzzling monstrosities the world has ever seen (SUVs), plans to drill the Arctic National Wildlife Refuge, the highest oil prices ever, and a corrupt Department of Energy and Patent Office that suppress promising new alternatives. Meanwhile both Mr. Bush and our vice president would like to see the largest peacetime military budget ever, and both support robust trade agreements that will benefit the rich and plunge the Third World into ever more environmental wreckage along with sweatshop/child labor conditions—all to give them more campaign donations and better prices at WalMart.

We have as never before an opportunity to elect someone who would truly support the cause of clean energy: Ralph Nader. Not to vote for this man is a vote against clean energy. Not to vote for Nader is a vote for business as usual. I strenuously object to the defensive posture, "A vote for Nader is a vote for Bush". Besides in some swing states, we will get nowhere with such defensiveness; that is media nonsense!. Many a nation has been doomed by such tentativeness. Only Nader will be able to lead the way to clean energy. He is the only candidate who has repeatedly declared the clean energy goal, one I am sure each of us here agrees to.

A vote for Nader is a vote for hydrogen, pure and simple. If he doesn't win this time, he could make a strong enough showing to win next time, perhaps after taking our turn with the oilmen in charge, who would show their true colors as to what not to do. I'm emphatic and unrelenting on this point. The media and polls and shallow pundits are to be vehemently distrusted in ignoring the Nader factor and in creating the illusion of a two-way race, consisting of personalities and non-issues. My colleagues abroad are appalled at all this! Only our consciences in the polls will decide the day: let's not forget that.

And now for the easier part, the technical aspects of the hydrogen economy. I see hydrogen as one of several key elements to a near-term victory over petroleum and other forms of dirty energy which must be wiped off the face of the planet (perhaps with the exception of some petrochemicals). I'm sorry for my strong language, but its time is long overdue. I refer to my new book for details on this one.

As for hydrogen, I have the following view: I see it as an interim fuel on the way to true "free energy", devices that could produce electricity from the zero-point field. I see also hydrogen coming from cold fusion and other devices that could produce the hydrogen from energy-efficient electrolysis to be used in fuel cells and hydrogen gas cells. Any seminar, conference or congress addressing the hydrogen question must look at these longer-term applications as well as the nearer-term transitional technologies. So let's bring in the likes of Gene Mallove, John Bockris, Stanley Pons, Martin Fleischmann, Randell Mills and other pioneers familiar with both the electrolysis and cold-fusion aspects of research leading to practical hydrogen engines, where the problem of energy-intensive hydrogen production and storage could be solved. We need to understand, for example, a safe and proper hydrogen gas tank pressure, materials and configuration. We want to know the optimal infrastructure.

From what I can tell from the literature you sent me, we all begin with a baseline concept for transitioning to a hydrogen economy, based mostly on the innovative thinking of Mr. Roy McAlister. I see many attractive qualities to that concept. But it is easy to create a "bandwagon effect" around a particular concept, later to discover unexpected flaws that cut into the optimism expressed, for example, in the fact sheet circulated before this conference. I throw caution to the wind. If I were asked the question, if tomorrow would we go for a continuing petroleum economy or a fledgling hydrogen economy (and we had only those two choices) I would enthusiastically vote for hydrogen. But I'd like to look at all alternatives and for optimal mixes, which means leaving some of our preconceived advocacies at the door. (perhaps including even mine pertaining to Nader) I also want to leave at the doorstep any proprietary patent agreements. This issue is so important that the world cannot afford industrial secrecy. We will have to find other ways to reward the innovators. I warned you I wouldn't be too popular, but we have a civilization to save as a higher priority than some capatalists' economic self-interest.

I'd also like to learn about how hydrogen fuel could replace existing fossil fuels in power plants. For example, I have been invited to address a group of angry citizens in San Diego, whose power rates over the past year have quintupled. They want rooftop solar collectors now, a fuel cell in their homes now, or perhaps a fuel for their powerplants— anything that would drop their rates. They're hopping mad. How can our group interface with them? We already have customers if we could gear up and tailor our products to them in time. While careful study is important, time is of the essense in the real world, and our consensus as to a sensible path into the future will be essential.

I'll give you an example from quite a while ago about how an engineering project concept changed rapidly under duress and for the better. I was on one of the planning groups for the Apollo program. Men of the stature of Roy McAlister convinced President John F. Kennedy we could go to the Moon. Others were unconvinced. Two schools of thought came out of this: one was the direct brute force method, Earth to Moon, and back with the upper stage. The other was Earth orbit rendezvous with a lunar-designed vehicle taking the astronauts the rest of the way and then back in Earth orbit to rendezvous with the Earthward craft. Earth-launch would use that ubiquitous, wonderful fuel, hydrogen, by the way, with life-giving fuel cells on board, a good metaphor of what we'd like to do now on Earth. No matter how hard we tried, no matter how many debates between the schools, neither of our tiger-teams could come up with an answer. Then, in a quiet corner of the room, one engineer finally had the floor. He said, what about lunar orbit rendezvous? We laughed at first, thinking such a suggestion to be ridiculous, but slowly we recognized it made sense. Using his approach, the mission could not only be achieved safely, but within schedule and budget!

I see our task to be similar. Ask not what hydrogen can do for us, ask what we can do to end this nightmare of a fossil-fuel economy asap, I mean now! Let's go beyond hydrogen and also look at ethanol and water, at electric motors, at hybrids, at emerging technologies of fuel cells, cold fusion, zero point energy, the latest solar photovoltaic technology, at everything reasonable—no stones unturned. My hunch is that hydrogen will be an integral part of this clean-fuel economy, but we shouldn't restrict ourselves to preconceived baseline scenarios. And we must make sure things work and are thoroughly tested before we move ahead. We are going to have to be very clever in our strategies, and with the same sense of urgency as if the world depended on what we come up with. So I believe we should set as a goal the end of the fossil-fuel age within the decade. We have the technology and R&D talent to create a clean and sustainable future for all time. Let's get on with it!

Incidents of Suppressing Discussions about the Suppression of Science

My (Almost Cancelled) Appearance at the 1999 Society for Scientific Exploration Meeting

In the summer of 1998 Professor John Bockris, program chairman of the June 1999 annual meeting of the Society for Scientific Exploration (SSE), invited me to give a keynote presentation on the suppression of science at that gathering in Albuquerque, New Mexico. I was honored by the invitation and gladly accepted the opportunity to share with some of the most prestigious new scientists in the world. I had been a long-standing member of this elite group of about 300 scientists, most of whom are university professors with tenure, and many of whom are cited in this book. Bockris himself is Distinguished Professor emeritus of chemistry at Texas A & M University, and has pioneered high-quality work on the hydrogen economy, and cold fusion.

Then, about five months after the invitation and six months before the event, I received a registered letter from Bockris informing me that the executive council of the Society went on "disconnect" when my name came up as a speaker, that I was no longer a scientist but a writer. Therefore I was dis-invited, even if I paid my own way there. Their decision may have also had something to do with my advocacy of inquiry into the "face" and other strange features on the Cydonia Plain of Mars— something felt to be nonsense by some of the council members who hadn't done any research on the matter. I was defrocked even by my new colleagues because I was viewed as being "controversial".

The irony of all this was that the opportunity to talk about suppression has been suppressed by those who purport to want to know more about the subject. Bockris was shocked with the council's decision, as they usually rubber-stamp the speakers listed by the program chairman. Bockris himself is no stranger to suppression. Once a group of reporters from Science magazine, the journal of the American Association for the Advancement of Science (AAAS), came to Bockris' laboratory to interview him on his experiments on low-energy nuclear transmutations, which can also be called alchemy. Because of a disbelief that such a thing could happen at room temperature, the reporters accused Bockris of fraud.

In the end, Bockris was vindicated by the thorough and high quality of the presentation of his important work. On another occasion, he was to host a conference for cold-fusion scientists in his chemistry department, but his own colleagues at Texas A & M voted unanimously to not allow this "pseudo-science" on the campus. The conference was instead held at the College Station Holiday Inn. These kinds of suppression are very common. During my presentations I'm in hotels twenty times more often than at universities and government facilities combined. Bockris really deserves a Nobel Prize, not the Ignoble Prize which the opinionated and uninformed skeptics would prefer to offer.

I protested to the SSE its decision to dis-invite me. To the credit of the SSE leadership, my invitation was reinstated after I reminded them of my active membership and a good record as a scientist. So new scientists need to be resilient and persistent in standing up for what is just, or just discarding it as petty (they recently demoted me from full membership so I've opted for "emeritus member"—retired from this culture).

The SSE meeting turned out to be stimulating and cordial. The papers presented were outstanding. We heard from leading researchers in cold fusion, parapsychology, zero-point energy, homeopathy, water restructuring, quantum physics, relativity and biology. It reminded me how much I miss mixing with colleagues whose work will certainly provide a foundation for the future, in this case towards a new science of consciousness. The abstract of my (unsuppressed in the end) talk at the 1999 Albuquerque meeting of the SSE follows:

The Suppression (and Resurrection) of New Science

copyright 1999, Brian O'Leary, Ph.D.
Abstract for address to the Society for Scientific Exploration,
Albuquerque, June 3-5, 1999

Research on anomalies suggests that we are in the midst of a major and unprecedented scientific revolution, based on principles of consciousness and contact.   Yet universities, industry, governmental agencies and foundations continue to resist support for most research on significant new findings.   The media fancy sensationalistically polarizing debates between overinterpreters of anomalous data and uninformed skeptics, thus bewildering the public. This situation is consistent with the historical studies of Kuhn and others on paradigm shifts.

Scientific disciplines are fragmented into specialties to the point where peer and funding pressures limit the scope of inquiry and lead to denials about realities beyond.   Because of this (often unconscious) conservative bias, scientists can become very unscientific when it comes to inquiry outside their own box.   They ignore the anomalies and blindly rely on existing theories, thus unwittingly helping perpetuate powerful vested interests and possible conspiracies.

New science is also fragmented into boxes. Researchers in psychokinesis, ufology, alternative medicine, materials anomalies, new energy and planetary anomalies rarely speak with one another, for fear that one's already shaky credibility would be further compromised by venturing into "fringe" areas outside his or her own specialty.

All this can create great anxiety for new scientists. Yet the ecological mandate and our thirst for the truth require that more scientists become less afraid of free and open inquiry outside the box. New paradigm scientists will need to be supported and respected for their work, not disdained and separated from old and new colleagues, as is now the case. We have the prosperity to create the needed infrastructure. The Society for Scientific Exploration could play a major role in bringing in the new paradigm.

My (Almost Cancelled) 1998 Commencement Appearance at the California Institute of Technology (Caltech)

Another story concerned an invitation I received from a graduating physics honors student at Caltech to speak during their Commencement weekend. He wanted me to speak about new energy, new science and their suppression—subjects I know very well and speak and write about all the time.

The Caltech connection came from the 1970-71 academic year, when I had been a visiting faculty member there while on sabbatical leave from my position as assistant professor of astronomy at Cornell University alongside the late Carl Sagan. My purpose that year was to work as deputy team leader during the planning phases of imaging science on the Mariner 10 Venus-Mercury flyby. The team leader I assisted was a Caltech professor of geological and planetary sciences. After the mission in 1975, our team was awarded a NASA Distinguished Group Achievement Award. I predicted and named the largest prominent feature on Mercury, the Caloris Basin. My year at Caltech was scientifically very creative and rewarding for me. I also published a number of papers in such prestigious journals as *Science*, using Caltech as my affiliation. Around that time the American Association for the Advancement of Science bestowed upon me the honor of being a Fellow of the AAAS. I'm sure there are those who now regret having done it.

When my former colleague, the Caltech professor, heard about the fact that I would be coming there to speak on the campus after a 27-year absence, he had a fit. He told the student who invited me to reverse the invitation. He questioned my genuine credentials and denied that I was ever at Caltech! His objections, I'm sure, came from my advocacy of inquiry into the Martian Cydonia features(chap. 6, ref. 1 and 3). Also my advocacy of cold-fusion research ran counter to Caltech's party line, presented in Chapter 2.

Again I protested and prevailed. But my presentation was moved to the end of the weekend, when most of the students were gone, so the attendance was low.

Polls Show That the Most Prestigious Scientists Are Biased Against Psychic Phenomena

My experiences with the elite are consistent with Dean Radin's look at polls taken of various groups of scientists, reported in his recent book *The Conscious Universe*. (Chapter 6, ref. 11). When the question was asked, do you believe in the certainty or possibility of psychic phenomena, 68 per cent of the public answered yes, 55 per cent of professors from a variety of disciplines, only 30 per cent of professional scientist-members of the AAAS, and a paltry six per cent of members of the elite National Academy of Sciences. From this, Radin concludes, "...if Joe Sixpack and Dr. Scientist both witness a remarkable feat of clairvoyance, we can predict that later, when we ask Joe what he saw, he will describe the incident in matter-of-fact terms. In contrast, when we ask Dr. Scientist what he saw, he may become angry or confused, or deny having seen anything at all."

So it's no wonder young people aren't learning much about new science. It has been censored at the academy. There seems to be an inverse relationship between the power and prestige of scientists and their willingness to look at a greater truth. They have too much of an investment in existing theories. Therefore, new scientists need to be vigilant, confrontive and unrelenting. There is a big paradigm battle going on now, but most of it is happening outside the hallowed halls of ivy.

APPENDIX IV

Towards a Quantitative
Understanding of Consciousness

In this appendix, I attempt to place numbers on some of the experiments using random event generators (REGs) described in Chapter 6. Analogous to the flipping of a coin billions of times, one would expect that the mean probability to be fifty per cent in a random process (no intention). In their data base embracing thousands of data-rich trials of dozens of operators, Jahn and his collaborators were getting numbers like 50.02 per cent in positive intention (analogous to correctly guessing "heads" in a coin flip) and 50.03 per cent in negative intention (correctly guessing"tails"). (Chap. 6, ref. 9)

In 1996 I began to look at the concept of measuring the effectiveness of consciousness, based on these quantitative results. I presented this to the 1996 International Symposium on New Energy in Denver and at the 1997 International Symposium on Consciousness, New Medicine and New Energy in Tokyo. In these papers, I defined a "consciousness unit" as that percentage of a human-machine interaction attributable to the consciousness of the operator (the non-random component). In Jahn's experiments, the mean value of the effectiveness of all operators was therefore about .05 per cent, or .05 consciousness units, reflecting the spread of .02 per cent in the positive direction, and .03 percent in the negative direction. The effects amplify about six times for bonded couples, or about 0.3 consciousness units.

The late Bruce and John Klingbiel of Spindrift, Inc. also did some experiments showing how powerful consciousness can be. (Chap. 6, ref.

2) They revealed some cases of operator-machine combinations that can attain as much as ten to twenty consciousness units. Here the effects of consciousness grew for two reasons. First, the Klingbiels reformatted the stream of data to sidestep a hypothetical "unconscious saboteur" that tends to even out some non-random information in the data. This sabotaging effect could be greatly reduced by mathematical filtering techniques. The second amplification came about when the operator surrendered himself or herself to a higher will, as in the process of prayer or ritual.

More recent work by Dean Radin (Chap. 6, ref.11) and Roger Nelson at Princeton (Chap.6, ref. 10) show that groups gathered for prayer or humor can produce several consciousness units in a field REG. These peak experiences can also be produced during times when over a billion people are focussed on one event such as the Academy Awards, the World Cup, or the September 11, 2001 terrorist attacks.

We can anticipate that a major purpose of future experiments in consciousness is to find ways of increasing the psychokinetic (nonrandom) portion of the action. An ideal performance would be 100 consciousness units, or 100 percent effectiveness in influencing the results of a random event generator. This pure regime of consciousness with a binary numerical system can be likened to pure manifestation in the material world, such as would come from Sai Baba or Thomaz Green Morton. I described these "miracles" in *The Second Coming of Science*. This special state of being can be compared to over-unity power in a free-energy experiment. How can the results of both classes of experiments be quantitatively related, since we appear to be dealing with similar mechanisms?

Measuring outputs from both experiments will enable us to better understand the mechanics of consciousness in hybrid free-energy/consciousness experiments. Perhaps then we can better grasp what is meant by conscious intention, as if our minds were acting like solid-state free-energy devices resonating with the zero point field. We could then compare the over-unity power effect with amplifying consciousness "hits" to the degree that the consciousness component dominates the interaction and a positive feedback loop can be attained.

In addition to the counterintuitive results of quantum mechanics, two apparently disparate experimental approaches (free-energy device design and parapsychology experiments) appear to involve similar kinds of interactions:

- accelerating an electromagnetic charge in the ZPF can produce both free energy and psychokinetic "hits".

- the interaction can be "revved up" to a critical threshold in which either over-unity power is achieved or a sufficient number of consciousness units are attained, wherein significant and repeatable physical or energetic manifestations from the ZPF become possible. The recent experiments of university scientists reporting in the peer-reviewed literature clearly show that it is now possible to monitor the effectiveness of group consciousness and focus by using field random event generators.

- quantitative experimental results in both classes of experiments could lead to a mathematical theory of consciousness and the zero-point energy, where consciousness units can be related to free-energy units.

- the concept of mass dissolves into energetic interactions of accelerating charges with ZPF. These reciprocating interactions could form the basis of a new science of consciousness in which the mechanistic aspects are mere limiting cases of reality.

The preceding discussion suggests that both types of experiments (psychokinesis and free-energy production) involve accelerating a charge in the ZPF (either by intention or by designing a device) in which electrons are manifested from an imaginal (invisible) realm. The question becomes, how many electrons per second are made manifest by 0.01, 0.1, 1, 10 or 100 consciousness units? Is that relationship a linear one as in experiments on the mass equivalence of energy or the mechanical equivalence of heat? One scientist, Shiuji Inomata, has provided such a mathematical formulation involving both real and imaginary components of electromagnetic charge. (Chap. 6, ref.3) We can imagine that perhaps some day the power of intention could be as measurable as anomalous outputs in energy, as expressed in watts.

In Jahn's and Radin's random event generator experiments, the following thought exercise might help. Imagine new electrons being produced to cause an experiment to yield measurable consciousness units.

The challenge is to understand the functional relationship between anomalous energy output and consciousness output which will in turn give us more insight on the nature of consciousness and of free energy.

In the classic text *Margins of Reality*, Jahn and Dunne explore the meaning of consciousness as a function of space, time, charge, energy, frequency, wavelength and other physical parameters. (Chap. 6, ref.9) Using quantum theory and information theory as metaphor, they reluctantly conclude that only qualitative insight seems to be available, because the operator-environmental interactions are very complex. Perhaps some of the answers they are seeking could come from hybrid consciousness-free energy experiments, where electrical inputs are varied for free-energy experiments and electrical outputs are measured in both free-energy and consciousness experiments. In this way, the requirements for consciousness inputs can be better understood.

Perhaps then we may begin to determine what thresholds we will need to tap both the zero-point energy and psychokinetic powers. In relating the two classes of experiments, it should be possible to employ them together, using mathematical filtering techniques to enhance psychokinesis—and therefore free energy. In other words, we can create and utilize any number of electrons we wish for whatever purpose simply by focusing our energy on an appropriate device which can amplify the power of our intention. For example, we could tune a psychokinetic garage door opener to our psychic "frequency" and open the door simply by thinking about doing so. The great potential of these technologies also means that we must use them wisely for the benefit of the Earth and humankind, as described in Part II of this book.

APPENDIX V.

Opening to New Hydrogen Energy Technologies

Brian O'Leary, Ph.D.
Prepared for the World Congress for a Hydrogen Economy,
November, 2001

Abstract

When we talk about a hydrogen economy, we generally refer to the generation of energy from the combustion of hydrogen or from its conversion to electricity through fuel cells. These approaches have been with us for well over a century and have wide appeal because of their nonpolluting nature and cost-effective operations in the long run, when compared to a fossil fuel economy. But often missing from the discussion are more advanced technologies which could make a hydrogen economy all the more appealing and avoid the large expense of creating energy-intensive hydrogen production facilities and infrastructure.

Basically, the traditional hydrogen economy involves the exothermic reaction between hydrogen fuel and oxygen in the atmosphere. This chemistry is well understood. But two other approaches now being researched are much more promising and provide far greater amounts of energy.

One is the hydrogen gas cell and hydrogen plasma cell technologies of Dr. Randell Mills of BlackLight Power Incorporated. Early research indicates that, when hydrogen gas is heated in the presence of certain catalysts, the hydrogen atom appears to "shrink" to a state lower than the normal ground state. The result is a novel chemical state of hydrogen called a hydrino, and the release of energy approximately 100 times that of normal hydrogen combustion.

The second advanced technology concerns the ability for hydrogen to electrochemically, ultrasonicly or otherwise transmute into helium in a fusion reaction at room temperature, with the release of hundreds of thousands times greater energy than normal hydrogen combustion. Radioactivity is negligible. This process is often called "cold fusion" or "low energy nuclear reactions". It was first discovered in 1989 by Drs. Martin Fleischmann and Stanley Pons in an experiment using palladium cathodes in a solution containing heavy water. These experiments have been replicated many times and comprise an impressive scientific literature.

This paper will describe how a hydrogen economy could be enhanced by either or both of these important research developments. The result could become what the Japanese call "new hydrogen energy", where ordinary water becomes the clean-burning fuel of choice.

Introduction

In conferences such as this, we see a growing consensus that humanity cannot continue to go on with a fossil fuel economy. Dwindling supplies and the great pollution caused by burning oil, coal and natural gas necessitate a major overhaul of our energy systems. Hydrogen is clearly the energy carrier of choice because it is clean, safe, feasible, inexhaustible, and cost-effective in the long run.

The sciences of hydrogen combustion and hydrogen fuel cells are well-known and have been with us for a very long time. Henry Cavandish discovered hydrogen in 1766 by pouring acid on iron and collecting the bubbles it gave off. The gas in the bubbles was found to be combustible, later identified as hydrogen. In 1820, Reverend W. Cecil built an internal combustion engine fueled by hydrogen. Later in the 1800s and during the 1900s, hydrogen engines became increasingly refined to the point that they have become operationally competitive with petroleum-fueled internal combustion engines. [1] Ironically, the use of hydrogen fuels predated the use of oil, which later took over with the invention of the carburator and because of ease of production, storage and fueling.

Hydrogen fuel cells are also playing an increasing role in electrical power generation for transportation and buildings. Sir William R. Grove built the first hydrogen fuel cell in 1839. As in the case of internal combustion, fuel cells have been continually perfected to the point where they can produce electricity with about 90 per cent efficiency. The cost of fuel cells keeps decreasing as they become ever more common in the marketplace.

As promising as these traditional hydrogen energy technologies may be, we come up against fundamental constraints which chemistry places on how much energy we can get out of the hydrogen atom. Otherwise, the conventional wisdom holds that we must resort to thermonuclear explosions or multibillion dollar attempts to control a hot fusion reaction in a so-far unsuccessful effort to reach energy "breakeven" in a confined plasma.

When we rely on energy coming from combining hydrogen and oxygen chemically, there is a limit which dictates that more energy is needed to produce the hydrogen in the first place. A renewable solar-hydrogen and wind-hydrogen economy are viable answers to this problem. [2] Another approach is to seek a role which might be played by "new hydrogen energy" based on recent discoveries of powerful excess energies in experiments which haven't yet entered the realm of engineering or commercial application. Because we are are such a crossroads regarding our energy future, perhaps we can now begin to agree on designing a non-fossil-fuel infrastructure while exploring every available option. This paper reviews the most promising possibilities in advanced hydrogen technologies, which are admittedly still in the research phase of a research and development cycle.

The Discovery of New Hydrogen Energy

During the late 1800s science fiction writer Jules Verne predicted that abundant energy could come from water as a fuel. One century later, his vision seems to be fulfilled.

In 1989, University of Utah electrochemists Martin Fleischmann and Stanley Pons announced a dramatic discovery: deuterium in heavy water placed within the lattice of a special palladium cathode somehow fused with itself to produce striking excess heat and a hint of fusion products such as tritium and helium, but no harmful radioactivity. These often-replicated results could not be explained by traditional chemistry. Energy outputs of up to hundreds of thousands of times greater than inputs were observed, some cells operated with excess energies for up to two months, and the products of nuclear transmutations were commonly detected. In spite of repeated efforts on the part of mainstream nuclear physicists to debunk these results, obtained totally outside their expertise, many experiments showing similar effects have been published by competent electrochemists, ultrasonic researchers and others at the Electric Power Research Institute, the Stanford Research Institute, Los Alamos National Laboratory, Oak Ridge National Laboratory, Naval Weapons Center at China Lake, Texas A&M University, Hokkaido National University, and several other prestigious institutions. [3]

Then, in the late 1990s, Dr. Randell Mills of BlackLight Power, Inc., made a second astounding discovery. When he placed water in contact with a potassium- or other metal-compound catalyst, he measured an unexpected release of energy, apperently including as a byproduct a special fractional lower (collapsed) quantum state of hydrogen he called a hydrino. This novel hydrogen chemistry produced over 100 times greater energy than that from ordinary hydrogen combustion. He has constructed prototype hydrogen plasma cells coupled with a gyrotron to directly produce electricity. He claims to be able to scale up this apparatus linearly to generate any amount of power at less than one cent per kilowatt-hour, lower than any known alternative. [4]

These claims, while extraordinary, are well-grounded in hundreds of experiments which involve the principle of loading hydrogen into a metal lattice which somehow catalyzes the hydrogen into a collapsed state and/or fusion with itself, with the release of large amounts of energy. Some scientists are just beginning to understand this process from a theoretical perspective, but there is as yet no overall agreement on the underlying physics.[3,4] What makes matters even more challenging is that replication and repeatable results are difficult to achieve at this point. While the excess energy measurments and reaction byproducts are very real, the messy chemistry of catalyzing hydrogen with metals is still an inexact science. But should that cause us to abandon the overall effort? Of course not: the results have been clearly robust enough to continue the research and eventually to perfect the process. Continuing efforts will also help create a new theory of hydrogen chemistry and materials science. This all is part of the sciencific process of discovering, experimenting, modeling, researching, replicating—and later, engineering.

Some years ago, the Japanese government called this anomalous energy "new hydrogen energy", and they began an engineering program during the 1990s to attempt to produce a commercially viable model. As, we shall see, that project was to be doomed primarily because they claimed they were unable to replicate the process. They didn't seem to understand its complexity and they denied positive results reported by visiting American researchers familiar with the process. The phenomena were still elusive enough to remain in the province of basic scientific research, an issue we'll explore later in this article. The fact is, new hydrogen energy is still viable but not quite ready for the marketplace.

We might compare new hydrogen energy to the early days of internal combustion engines, fuel cells or any other remarkable new technology. We know these devices can work, but the time for building completely reliable commercial prototypes has not quite come yet, because we are still within the realm of science and not engineering. And yet the latter community tends to ignore very real progress. Perhaps for convenience, they chose the safe path of siding with the unknowing scientistic skeptics that attempt to deny the evidence. These debunkers often have prestigious credentials in other fields and have clout in the scientific community and the media. This sudden ending of the program need not have happened with a coordinated research and development effort among teams of scientists and engineers freely exchanging their ideas.

Because of its great promise, to ignore new hydrogen energy at this time might be likened to ignoring the potential of airplanes during the time of the first Wright Brothers flights. People were still focused on surface transport and slow airship infrastructures, which were later overtaken by the explosive growth of aviation. I believe that a similar scenario awaits us with new hydrogen energy.

Differences between the Realms of Science and Engineering

When I first entered the astronaut program in 1967 and later became a scientist on unmanned planetary missions during the early 1970s, I was surprised to discover that NASA was deeply divided in its engineering and science functions. [5] While the engineers were justifiably focused on the familiar tried-and-true technologies, the scientists were continually "pushing the edge of the envelope" on what was possible— whether it was state-of-the-art propulsion technologies or innovative instrumentation to explore the solar system. In NASA, the engineers were managers who usually won these debates and took the cautious path. Unfortunately, the trend in management philosophy is moving even further away from the scientific world view, as unknowing lawyers and business people take over the reins of leadership from engineers who at least might know a little bit about the technical issues.

This ideological divide holds to this very day, and no example is more poignant than that of new hydrogen energy. Not only is the greater community unaware of startling new developments and replications. The results continue to be senselessly debunked by mainstream hot fusion nuclear physicists who know little or nothing about electrochemistry or materials science, and who also feel they have a lot to lose if their multibillion dollar programs couldn't compete with the new results. Because the principles of new hydrogen energy appear to violate the nuclear physicists' understanding of theory, these irrational skeptics tend to hide within the theory and deny the obvious anomalous experimental results that challenge that theory. [3]

The result of all this is a premature public denial of the observed low energy reactions among hydrogen nuclei and atomic hydrogen collapse phenomena. The managers and engineers who will be designing the hydrogen economy will tend to be technologically conservative and will want to go with what has worked consistently for a long time. By default, perhaps, they join the debunkers, when their own abilities to strictly replicate results are fuzzy. The challenges of producing new hydrogen energy continue to be daunting and elusive, certainly not the kind of thing that could be replicated by reading a manual. But the day of rigorous replicability will almost certainly come with continuing basic research. This process of the chaotic, doubt-ridden, long but inexorable move from scientific discovery to engineering application has been with us since the time of Galieo. Especially during times of great change, people are afraid of the change and so deny its existence.

"Any sufficiently advanced technology is indistinguishable from magic", were the words of famed writer Arthur C. Clarke. He believes (as do I) that new hydrogen energy will have a 99 per cent chance of succeeding.

The Japanese New Hydrogen Energy Program

A vivid example of the rift between the scientific and engineering management cultures recently happened in Japanese efforts to establish a

$23 to $30 million new hydrogen energy program in 1993. The purpose of the program was to translate the astounding results of cold fusion science into engineering reality, but the project became a total failure and was shut down in 1998, because the engineers were supposedly unable to produce excess energy. Many cold fusion experts stopped by to participate, including Martin Fleischmann, Michael McKubre of the Stanford Research Institute, Ed Storms of the Los Alamos National Laboratories, and Melvin Miles of the China Lake Naval Warfare Center. All of these individuals are scientists whose positive results were denied by the Japanese group even though Miles produced excess energy to show to his colleagues right in the laboratory in Sapparo. Nobody wanted to look; their minds had already been made up. This behavior was reminiscent of Galileo's colleagues' refusal to look through his telescope.

In a carefully documented article, Jed Rothwell of *Infinite Energy* magazine believes that the Japanese program was killed by the incompetence of the engineers. [6] They had underestimated the difficulty of replication and were somehow not motivated to do the necessary scientific, creative and technical work that goes along with successful basic research. Whether by the conspiracy of vested interests among leading mainstream managers, scientists and politicians—or by simple ignorance about what to do and how to do it—this important Japanese effort died because of the lack of necessary teamwork. Their final report was entirely negative, inviting the outrage of the visiting scientists. It gave the public the impression that they tried, that they could not produce the needed effects, and that those reporting positive results were mistaken. Meanwhile the work of Miles and others continued to be ignored by his Japanese colleagues, perhaps because it would have been too much of a political loss for those in charge to change their minds so late in the effort.

New hydrogen energy now sits in an awkward void poised between research and development, between basic and applied science, between the tinkering, hard-to-replicate experiments funded on a shoestring and the creation of a commercial device, between proof-of-concept and the publicly-acknowledged industrial phase. It awaits that break into rigorous replicability. At this point, we are too far down on the toe of the profit curve to invite venture capital, but not so far down as to not invite altruistic capital. In spite of this, the anomalous energy effects continue to be robust and undeniable. The Japanese example becomes similar to the superficial U.S. Department of Energy Panel report denying cold fusion just months after it was discovered. New hydrogen energy is neither the stuff of establishment science nor of mainstream industrial engineering—yet.

Reconciling The World Views of New Hydrogen Energy Science and Traditional Hydrogen Science and Engineering

Those of us who are evaluating hydrogen futures need to become aware of the significant progress in new hydrogen energy research, which could revolutionize our energy picture in short order, once the necessary science and engineering are done. A sophisticated scenario for a hydrogen economy would include at least two eventualities: one is a "baseline" scenario which includes the tried-and-true hydrogen chemistry of internal combustion and fuel cells, as discussed in this Congress. The second brings into consideration new hydrogen energy and other new energy technologies, which could take the world by storm within a decade or two.

As we conceptualize and begin to build solar-hydrogen systems, we would also want to increase our support of new hydrogen energy research which could lead to commercially viable units within years. Why should we do this? We need to keep our options open until such time we can make a more rational decisions about our energy future. On the one hand, we want to move into an aggressive clean energy program using available technologies. On the other hand, we don't want to make an irrevocable commitment to a particular path which might be unnecessarily capital- or materials-intensive.

The funding of new hydrogen energy research would be two to three orders of magnitude less (on the order of tens of millions of dollars), and its development will be one order of magnitude less, than that for developing a solar-hydrogen infrastructure. The Mills technology, for example, would completely eliminate the need for any solar input at all. One gallon of water could run a vehicle for over 1000 miles without refueling. It could heat and provide electricity for a home for more than a year. Even greater efficiencies could be achieved with cold fusion and over-unity electromagnetic devices.

We also need to discuss which scenarios serve the public interest the most. Are the military applications of powerful new energy technologies sufficiently threatening to want to stop their development and to move instead towards a traditional hydrogen-solar economy? Or can we build in safeguards? These are important questions we must address as we move beyond the denials of new energy. We need to assess all viable clean and renewable options that lie ahead. New hydrogen energy would be far most economical and have least impact on the environment. Power could come from small cells deployed wherever they're needed. There would be no more use for large energy infrastructures, no grids, pipelines or production centers.

Hydrogen is amazing stuff. Our knowledge about its combustion and use in fuel cells have been with us since the birth of the Industrial Age about 200 years ago. The relevant technologies have been available for almost that long, and now they're being given a new look for fueling the 21st Century and beyond. This comprises our "baseline scenario" for a solar-hydrogen economy. But infrastructures would have to become more centralized, more expensive and consume more natural resources than a new hydrogen energy economy.

A second major hydrogen energy technology was born during the Modern Age, with the first explosion of a hydrogen bomb fifty years ago. We discovered how hydrogen could be used for destructive purposes. Since then, we have been struggling to develop hot fusion reactors to generate central-station electricity. The problems with this option are technical challenges, high capital costs, continuing grid systems, more centralization than ever, and some pollution and radioactivity.

It has only been within the past twelve years that a third set of revolutionary hydrogen technologies have emerged. The new hydrogen energy units of the Post-Modern Age will be small, decentralized, cheap and convenient. They will most likely satisfy both environmental and economic criteria for success in the future. We cannot ignore this fact at a time in which we as a culture must wean ourselves from fossil fuels. Engineers, managers, scientists and industrialists alike will need to gather to propose scenarios for consideration as to the best way to proceed in the public interest and not necessarily the interests of vested industry.

Fortunately, the rift between science and industrial engineering is above malice. They are simply different worldviews in which each culture doesn't adequately understand the other. Scientists and inventors tend to be unrealistic about their expectations of commercial application. Engineers tend to cut themselves off from the science prematurely because of their interest to implement a new future which is based on cut-and-dry technology. The shutdown of the Japanese effort to demonstrate new hydrogen energy is an example of this oversight. This kind of rift most often happens at the time of a bold new breakthrough. It can be healed by open discussion of how real and enduring these phenomena are, and how further investigation could almost certainly lead to an energy revolution.

There is actually more malice amongst factions of scientists—especially when the breakthroughs of one set of disciplines (e.g., electro-chemistry and materials science) overtake those of another (e.g., nuclear physics). The startling new discovery that there is an energy-producing relationship between hydrogen and some metals unseats the prevailing wisdom that hydrogen can only give off so much energy unless we attempt to confine it within a hot plasma. These squabbles can infect and spread to those who are not aware of the technical content of the debate. In fact, there are many approaches to new energy, any one of which could change the world. [3]

In truth, scientists creating breakthroughs and engineering managers are two cultures that could work effectively together, two sides of the same coin. They just need to talk with each other and compare their world views. The engineers need to know more about the dynamics of premature debunkery, and the scientists should be more integrated into management. We all want to be united in purpose, which is to end our dependence on polluting and scarce fossil fuels. With our ever-increasing knowledge of the mysteries of hydrogen science and engineering, we can move to a clean and everlasting hydrogen energy future.

References and Notes

1. An excellent survey of the traditional hydrogen economy is *Our Future is Hydrogen!* by Robert Siblerud, New Science Publications,Wellington, CO., 2001.

2. John O'M. Bockris and T. Nejat Veziroglu, *Solar Hydrogen Energy*, McDonald Optima, London, 1991.

3. The state-of-art of advanced hydrogen technologies is ably reported in the bimonthly magazine, *Infinite Energy*. New Energy is reviewed in Appendix I of this book.

4. Randell L. Mills, "BlackLight Power Technology: A New Clean Energy Source with the Potential for Direct Conversion to Electricity", presented at the International Conference on Global Warming and Energy Policy, Ft. Lauderdale FL, Nov. 2000, Global Foundation, Inc.; also www.blacklightpower.com.

5. Brian O'Leary, *The Making of an Ex-Astronaut*, Houghton Mifflin, 1970.

6. Jed Rothwell, "The Collapse of the (Japanese) NHE Project", *Infinite Energy*, Issue 30, 2000.

INDEX

What's happening to us and Our Precious Planet?

Do we ignore our responsibility and toss ourselves willy nilly into the winds of terrorism, escapism, cynicism, political games and economic tyranny or can we rise above the din and babble? *Re-Inheriting the Earth* brings us directly into common-sense solutions. Specifically, this book:

- Researches and reports the facts about the deteriorating physical state of the world through the watchful eyes of a veteran Ph.D. planetary scientist and author;

- Reviews in an expert and unbiased way the latest developments in clean, renewable and cost-effective energy (including on-site visits)--solar, wind, hydrogen, cold fusion and zero point energy, which could soon end the fossil fuel age;

- Gives specific proposals for reversing our abusive global industrial, forestry, agricultural, and water use through a mix of traditional and innovative approaches;

- Reveals the corruptions of and practical answers for culture steeped in pollution, greed, war and secrecy;

- Critically examines and synthesizes global problems and solutions from scientific, ecological, philosophical, political, social, spiritual and personal perspectives;

- Calls us to action as world citizens to provide the needed blueprints, options and strategies for a sustainable and just future for humanity;

- Examines the nature of our tyranny and looks at ways in which we can change socially and personally to meet the challenges, including new ideas about forming a global democracy/ republic that could really work--if designed properly;

- Explores our potential for personal and planetary healing through our intention in consciousness -- not merely as speculation but as scientific fact; and

- Provides hundreds of useful references, five scholarly appendices, and an index for the serious student of sustainability, social change and frontier science.

For your personally signed copy, send $23 (in U.S.) or $30 (outside U.S.) to Brian O'Leary, P.O. Box 27, Washington, CA 95986. For more information, consult www.independence.net/oleary.